ELECTROPLATING
FOR AMATEURS

ELECTROPLATING
FOR AMATEURS

J.A.Poyner

FOX CHAPEL
PUBLISHING

Contents

Publisher's Note

This classic edition of Electroplating is published as a historical reference and for informational purposes only. Many valuable insights can be gained from learning processes used in the past. However, these are not standard practices today and many may constitute a serious health or safety risk.

CHAPTER 1

Introduction and Principles of Electroplating

Present day electroplating has become a well-established branch of metal finishing. Electroplating is a multidiscipline of engineering, mechanical and electrical, in co-ordination with applied chemistry.

In the early days of electroplating the industry started with wooden vats, D.C. generators, experience and 'rule of thumb' methods of process control. Over the years new metal finishes have been introduced. Automatic plant has been developed to cope with the increased volume of parts to be finished and to control the process, ensuring a constant quality of finish. Increased uses of metal (steel, stainless steel) and various plastics have been seen in the making of equipment for the finishing shop.

Present day finishing shops offer a wide and varied range of finishing processes: ion and gas plating, high speed selective plating, anodizing and electroplating on aluminum. Various electroless finishes cover a wide range of engineering requirements. Various alloy platings are carried out, such as gold cobalt, which gives a hard thickness of gold. There is also brass plating for electroplating safety pins, and components which have various rubbers bonded to them.

Over the last fifteen years development has been carried out on the plating of plastics. The technology of printed boards in the electronics industry has added impetus to the development and many plastics can be successfully electroplated. With the various processes involved in electroplating and current requirements of health and safety, appropriate precautions must be undertaken to avoid accidents and reduce pollution of the environment. These are dealt with in one of the following chapters.

With most of the processes used in model engineering there is ready-made equipment sold on the market, obvious examples being lathes, milling and drilling machines, along with the materials, metals and plastics. In comparison, with the electroplating process there is very little choice available on the market, apart from kits for electroplating. This is due to their limited use as compared to the machining and fabricating operations in model engineering and in small workshops generally.

The other main reason, however, is the degree of availability of the chemicals. Certain chemicals are restricted, and restrictions are placed on them in transit. The electrical equipment needed can be adapted from other sources – electrical test equipment, Avometers, or battery chargers or large capacity electrical cells. If desired a permanent rig can be made. This is useful for a continuous volume of components that have to be finished. A wiring diagram is included in the chapter on the supply of current.

With regard to the tanks required, this is dependent on the size of the component to be electroplated. A useful size is the 5 liter plastic ice cream container. These are useful for most pre-treatment and electroplating solutions. For warm or hot solutions, ways and means of heating the solution may be considered, such as fish tank heaters, or, if using a stainless steel or mild steel tank, a gas ring or electric hot plate may be used.

Chemical glass beakers made of heat-resisting glass may be used and can be heated on an electric hot plate or over a Bunsen burner with a suitable stand and gauze. This equipment can be purchased at most laboratory equipment suppliers.

The model engineer must decide on what size and volume of components he wants to electroplate, and what finishes he wants to use. These points will have to be considered, whether he wants a rudimentary or a more substantial electroplating facility. The other relevant factors are the space available, cost, and the convenience of using the equipment. For example, considering one finish for similar size components and a steady volume, it would suffice to have a simple facility of an alkali cleaner, a pickle made of diluted acid, with a rinse tank containing cold water or preferably running water. It could be made even simpler for certain components by giving them a scour with abrasive powder, then rinsing in cold water.

After these pretreatments, the components are electroplated in whatever electrolyte is chosen.

For an electrical supply a 12 volt battery, or a battery charger of 12 volts or 6 volts, may be used.

At the other extreme, for varied components and large volume, one could use an elaborate line of pretreatment and rinse tanks, with a line of electroplating tanks all combined with the custom-built rectifiers, heaters and agitation. The cost of this would be considerable, and there would be the problem of disposing of effluent and spent chemicals.

PRINCIPLES OF ELECTROPLATING

The fundamental laws of electroplating are based on Faraday's two laws. These state:-

(1)　The weight of metal deposited is proportional to the quantity of electricity passed.
(b)　For the same quantity of electricity, the weight of metal deposited is proportional to its electro-chemical equivalent.

These two laws need a little explanation to understand their implications. This is best provided by defining the units. In law 1, the weight (w) is in grams or ounces and the quantity of electricity is in coulombs, which is amps (a) x time (seconds) (t).

Therefore w is proportional to a x t.

In law 2, the electro-chemical equivalent is defined as the weight an element will replace or combine with eight parts by weight of oxygen in a reaction.

The valency is defined as the number of atoms of hydrogen which will combine or replace one atom of an element, in this case a metal.

The relationship between atomic weight and electro-chemical equivalent is: atomic weight = electro-chemical equivalent x valency. An example is nickel, atomic weight = 58.71, valency 2. Therefore electro-chemical equivalent $= \frac{58.71}{2} = 29.35$.

The Faraday as a unit is defined as a quantity of electricity which is (amps x time), which will deposit 1 gram equivalent of the metal.

The figure for 1 Faraday is 96,500 (amps x time) or coulombs. Now 1 ampere hour is 1 amp x 3600 seconds, or 3600 coulombs. So in 1 ampere hour the weight of a metal deposited will be:

$$\frac{\text{electro-chemical equivalent x 3600}}{96,500}$$

For nickel this would be $\frac{29.35 \times 3600}{96,500}$

= 1.095 grams of nickel per ampere hour.

In the process of electroplating the amount of the metal deposited is usually less than the calculated amount. This is due to hydrogen being generated in the process. The difference between the calculated deposit (100% efficiency) and the actual deposit gives the cathode efficiency, which varies with the electrolyte formulation.

Below is a table of the commoner electro-deposited metals with the commoner electro-chemical equivalents and valencies.

Various other terms will be mentioned in parts of the book. At this stage, a list of terms and their definitions will help to understand the function of the process.

Anode
This is usually the part in the process that will deposit the metal on the component. A piece of the material, such as zinc or tin, will be suspended in the respective electrolyte with the component, and when the current flows, the metal will be deposited. *This is an electrode positively charged, denoted $\oplus$.*

Cathode
This is the part in the process that will receive a deposit of metal from the anode via the electrolyte when there is a flow of current. The cathode is the component to be electroplated. *This is an electrode negatively charged, denoted $\ominus$.*

Metal	Atomic Weight	Valency	Electro-chemical equivalent (E.C.E.) $\frac{\text{Atomic weight}}{\text{Valency}}$	Deposit per ampere hour $\frac{\text{E.C.E. x 3600}}{96,500}$
Zinc	65.40	2	32.70	1.22
Nickel	58.70	2	29.35	1.09
Tin	118.70	2	59.35	2.21
Silver	107.90	1	107.90	4.02
Copper	63.50	2	31.75	1.18
Cadmium	112.40	2	56.20	2.09
Iron	55.80	3	18.60	0.69

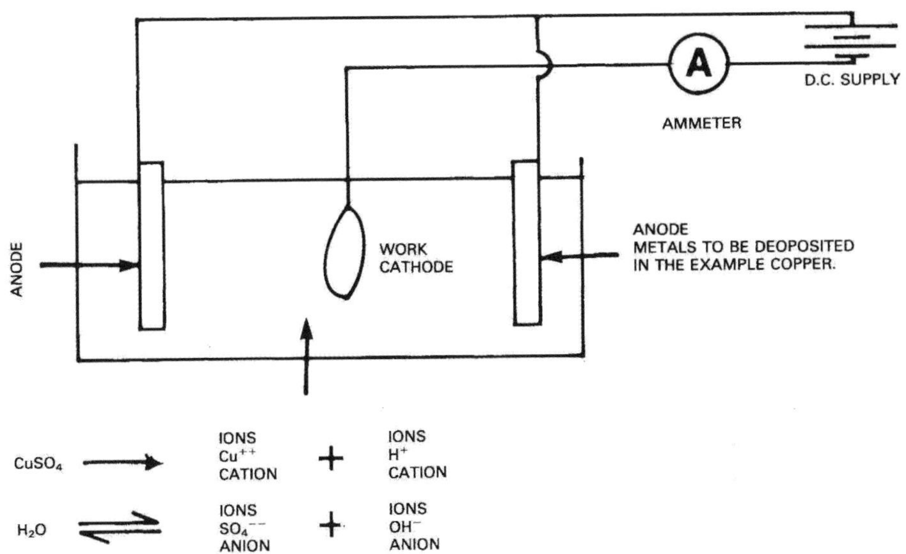

Fig. 1 *By definition the cations (Cu⁺⁺) are discharged at the cathode (work) as a deposit of copper, and also the hydrogen (H⁺). The anions (SO₄⁻⁻) and hydroxyl anions (OH⁻) are discharged at the anode, removing some of the copper anode (Cu) and forming copper sulphate (CuSO₄) which replenishes the concentration of copper sulphate in the electrolyte.*

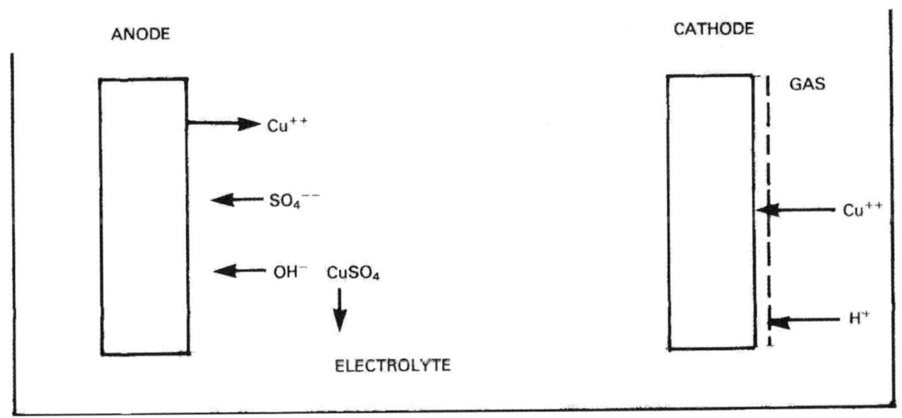

Fig. 2

Electrolyte

A conducting medium for most electroplating. An aqueous solution with water-soluble chemicals of the metal to be deposited. These chemicals dissolve in water and form ions which acquire a positive $\oplus$ or negative $\ominus$ charge.

Ion

Is an atom, or atoms, which have gained or lost electrons and in consequence carry a positive $\oplus$ or negative $\ominus$ charge. The positive charged ion is called a *cation,* which is discharged at the cathode. The negative charged ion is called an *anion,* which is discharged at the anode.

p.H

The use of the term p.H will be mentioned in the book in various chapters. The definition of the term p.H is defined as the log to the base 10 of the concentration of the hydrogen ion, or p.H $= -\log_{10}[H^+]$.

In electroplating it is used to define the acidity or alkalinity of an electrolyte.

In practical terms a p.H of 1 to 6 is acidic; p.H 7 is neutral; and p.H 8 to 14 is alkaline. A good example of neutral solution is pure water which is p.H 7.

Inert, or insoluble anode

This is when an anode such as stainless steel, platinized titanium or lead is used in an electrolyte and the anode does not dissolve into the electrolyte to keep the concentration in balance. This is as distinct from the copper anode in the copper electrolyte.

The inert anode has an advantage in certain electrolytes for electroplating. However, tighter control is needed in monitoring the balance of the electrolyte, due to lack of replenishment from the anode of that particular metal, which thus reduces the concentration of the metallic ion in solution, causing an imbalance in the electrolyte. The other factor affected by an inert anode is the p.H; this will change and affect the efficiency and physical characteristics of the electroplated deposit. The main electroplating processes using inert anodes are chromium electroplate (decorative and hard), using lead anodes, and gold electroplating, alkaline and acid electrolytes respectively.

CURRENT DENSITY

This term is defined as the amount of current (amps) per unit area of cathode (component), usually expressed as amounts of current (amps) per square foot or square decimeter, abbreviated *a.s.f. or a/dm^2.*

The first thing that must be known about the electroplating electrolyte is the preferred current density range. This varies with each electrolyte. The second consideration is the total surface area of the components you are going to electroplate. The third thing to consider is the shape of the component or components and its position in relation to the anodes in the tank.

(1) The range of current density will be given for the electrolytes in the chapter on electrolytes.

(2) The measurement of the surface area of the components to be electroplated require an elementary knowledge of mensuration. Some helpful examples for working out the surface areas are shown (Fig. 3). Most components are shaped in a combination of the listed shapes, or approximating to these shapes, so utilizing the appropriate formula

HEIGHT
LENGTH
RECTANGLE
AREA = LENGTH x HEIGHT
X2 WHEN ELECTROPLATING BOTH
SIDES
HEIGHT
LENGTH
TRIANGLE
AREA = HEIGHT x LENGTH
2
RADIUS
CIRCLE
AREA = π(RADIUS)²
CIRCUMFERENCE = 2π x RADIUS
DIAMETER
SPHERE
AREA = π(DIAMETER)²
SLOPE
DIAMETER
CONE
AREA = π x DIAMETERX SLOPE
2
RADIUS
HEIGHT
DIAMETER
CYLINDER
AREA = π x DIAMETER x HEIGHT

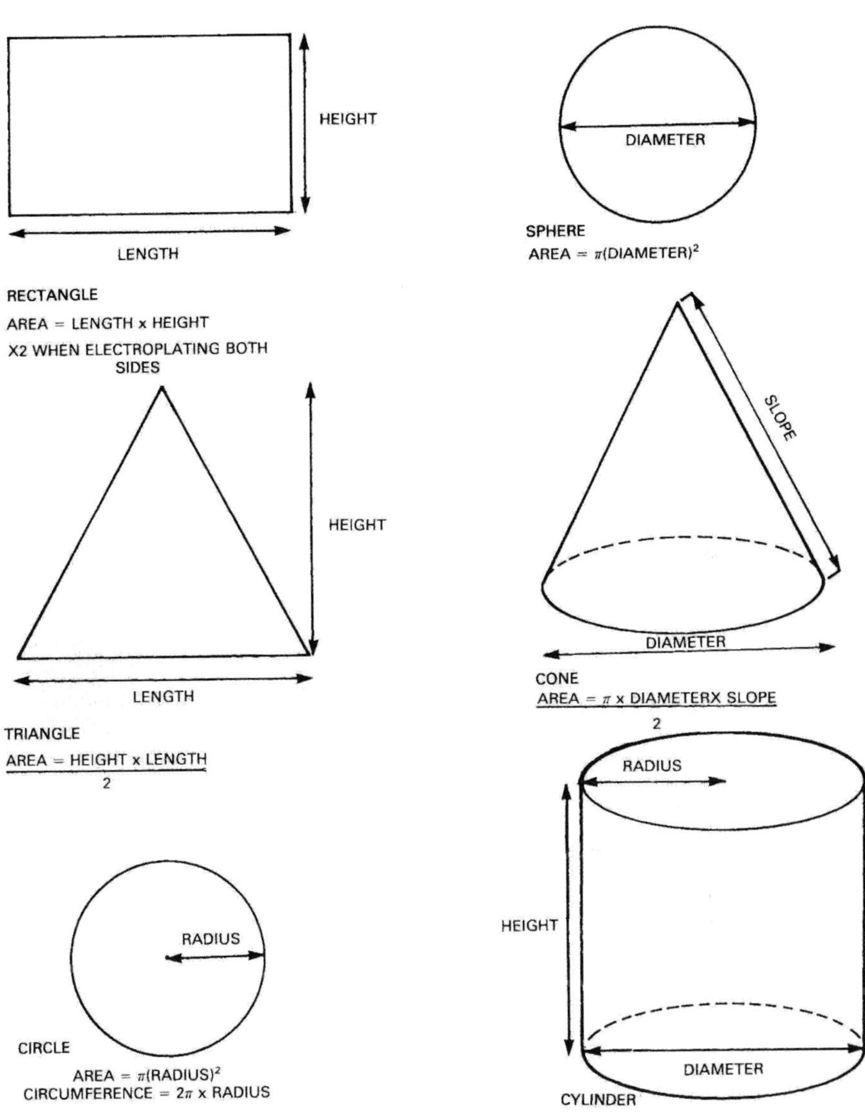

Fig. 3

a surface area can be obtained to achieve a good plated appearance with the requisite thickness. In other words, it is not absolutely necessary to be accurate to a fraction of the area to obtain the correct electroplated effect and thickness.

(3)   The shape of the component, or components, and position to the anodes in the electroplating tank can best be explained by a diagram.

Fig. 4 shows an electroplating tank with anodes and a cathode suspended in the electrolyte. The cathode is so shaped that it has areas labeled high current density and low current density. In electroplating this component within the range of the current density for the particular electrolyte the component may have a coarse, rough deposit burned in the high current density area and little or no deposit in the low current density area. To improve the deposit various means are possible. The anode in front of the high current density area may be moved to one side of the component, or the bottom high current density area may have a "robber" attached, which takes the excess current. For the low current density area the anode is bent, or a sub-anode attached nearer the area of the component. Another method is to reduce the current density to the lower part of the range, and increase the time for electroplating.

These methods are based on the art of electroplating, and with a build-up of experience they will not be too difficult to accomplish.

However, most components are of a straightforward shape and will electroplate quite well when suspended in a bath.

The other important shapes in electroplating are blind holes. These prove difficult, leaving stains around the hole. The best way to reduce this problem is to fill the hole with wax or similar material; this will in effect stop electroplating in the hole, but will reduce the staining effect on the component.

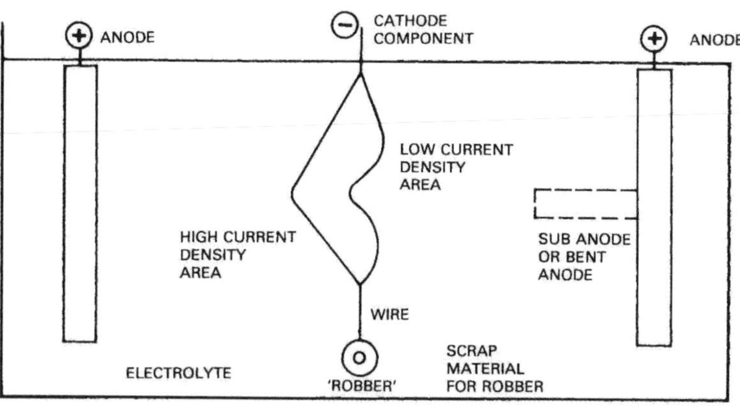

Fig. 4

CHAPTER 2

The Electrical Supply

Metals are mainly electrodeposited by the action of an electrical current. Direct current (DC) is essential for electrodeposition. Alternating current (AC) will not deposit metals. However, there are variations on alternating systems which are still being developed. These methods give a smoother and denser metal deposit and are used in certain specialized engineering applications.

The main consideration for an electrical supply in the workshop or light engineering facility will be derived from the normal single phase AC supply.

The supply voltage is immaterial, as a transformer is used to convert it to a working AC output.

The transformer should give the characteristics as follows:

CHOICE OF OUTPUT VOLTAGE

One main continuous winding tapped at 4,6,8,10 and 12 volts, capable of giving a substantial current of up to 10 amps continuous.

RECTIFICATION OF THE CURRENT OF THE AC SUPPLY

The function of the rectifier is to convert the AC current from the transformer to DC current. Rectifiers can be of the solid state variety, diodes etc., or the older copper oxide or selenium types.

As described in chapter one, Faraday's Law states that the mass of element, in this case metal deposited, is directly proportional to the quantity of electrical charge, coulombs or (amps x seconds). The rectifier should be suitable for rectifying 12 volts and passing 12 amps without heating effect, and be of the double wave type giving full wave rectification, thus giving a reasonably smooth DC output.

VARIABLE RESISTANCE

The next item to complete the package of electrical equipment is a variable resistance. This usually consists of a rotary switch set on a heat-resisting board. For high amperage the variable resistance board is often made of metal or slate. Set into the board are brass studs arranged in a circle, with a sliding contact with a handle made of brass. Arranged and connected behind the studs are coils of metallic wire, varying in size and shape to produce different resistances to obtain the desired amperage. The variable resistance is then connected up with an ammeter and voltmeter to permit convenient monitoring

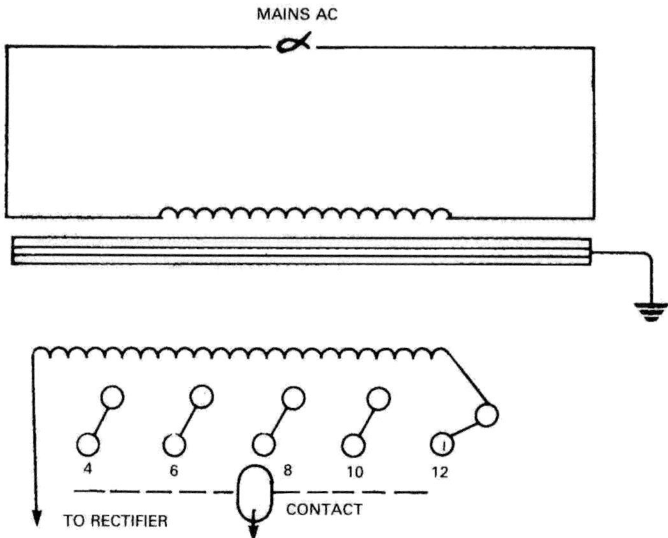

Fig. 5

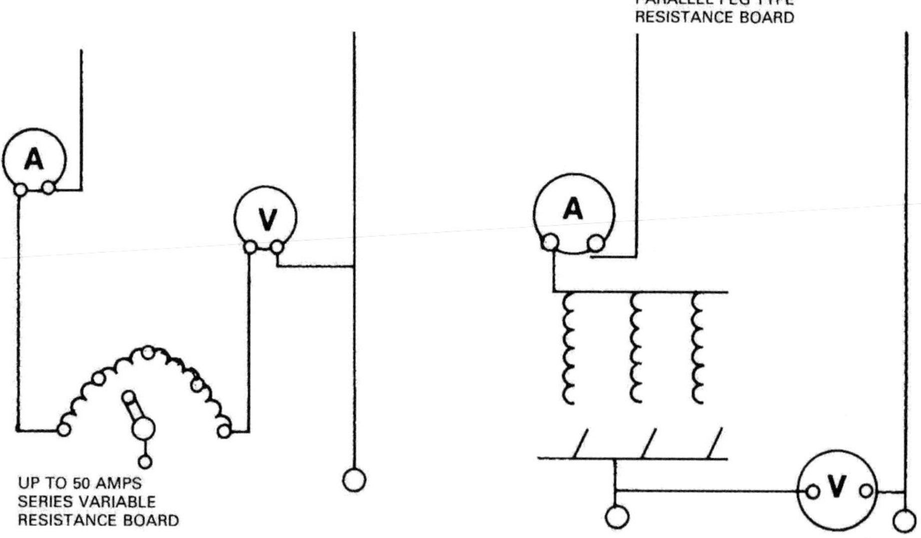

Fig. 6

Fig. 7

14

of the amperage and voltage readings. (Fig. 6).

On some modern boards, instead of a sliding contact on brass studs a resistance wire is used, with pegs and switches interspersed, which are pressed down for the desired amperage. This is when 30 amps or more are required, and by placing the coils in parallel instead of series, less heat is generated at the coils for the high amperage. (Fig. 7).

An ideal small electroplating electrical supply can be set up using a battery charger on its own, or better still connected up with a voltmeter, ammeter and variable resistance.

The home battery charger normally has only an ammeter, and sometimes a 6 to 12 volt plug, the maximum output being 4 to 5 amps. If you have a 6 volt output, use this for normal electroplating. A 6 volt control unit must be made as a separate item, either on its own chassis or in its own metal box. If made in its own box, do not forget to provide adequate ventilation, as quite a lot of heat is generated when using maximum amps.

The 12 volt unit can also be made as a

A typical layout of the electrical supply and an electroplating tank with an electrolyte (zinc), and zinc anodes and components being plated.

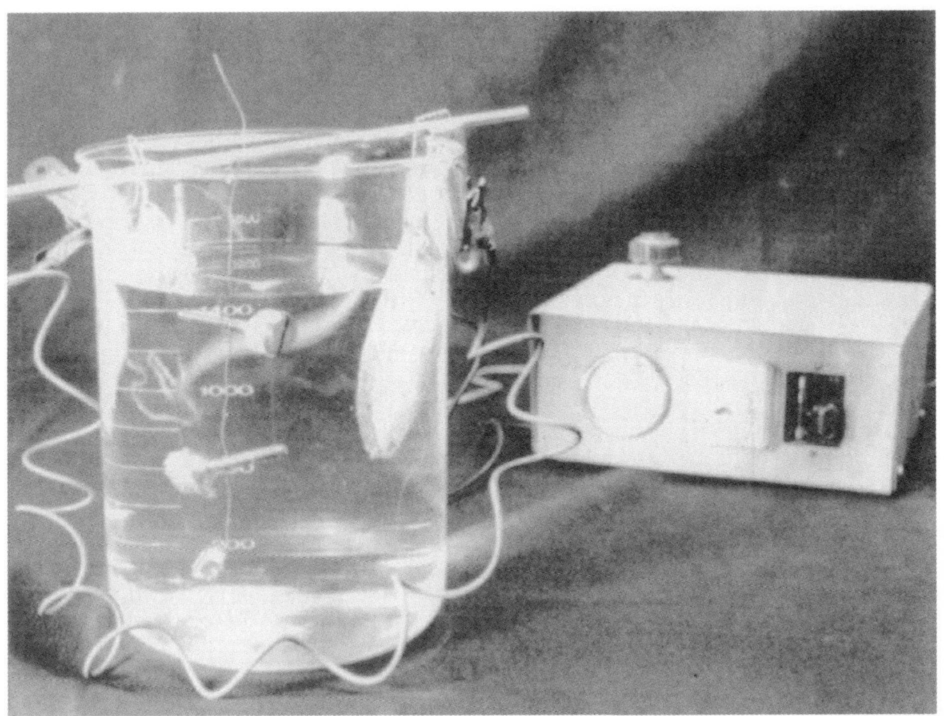

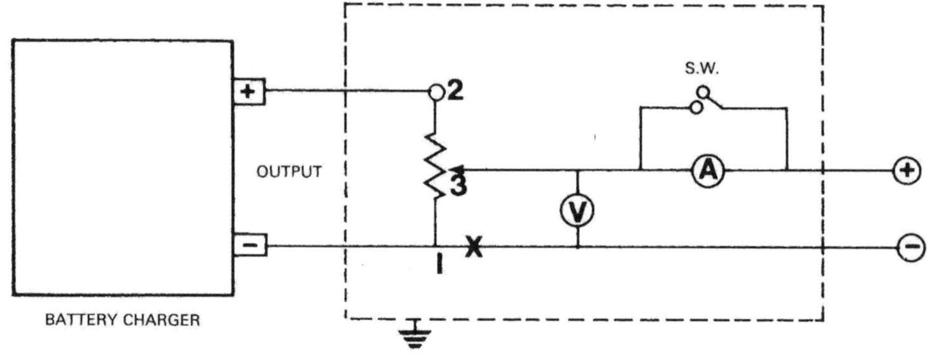

Fig. 8

separate item if desired, but if you have enough room inside the battery charger casing, you can fit the control items into this. Then, when the unit is needed as a battery charger, wind the potentiometer knob to maximum volts, making sure that the one amp meter is shorted out using the 5 amp switch, and you are ready to charge.

Should you make the control unit as a separate item, remember to connect the chassis or metal case to the metal case or chassis of the battery charger.

Fig. 8 shows the circuit diagram which is very simple.

The potentiometer is connected across the output of the battery charger, numbered 1 and 2 on the diagram, and the sliding arm 3 taps off the voltage as required, and is shown on the voltmeter V.

If you are electroplating a component that requires more than one amp, then the one amp ammeter must be shorted out by closing the 5 amp switch S.W. The higher amperage can then be read on the charger unit's internal ammeter. When you wish to electroplate a small compo-

nent requiring less than one amp, turn the potentiometer right down so as to put minimum voltage in the circuit, open the switch S.W., and with the anodes and components (cathodes) already in the electrolyte, connect the control unit output to the respective anode and cathode connections. Then wind up the potentiometer until the required reading on the one amp ammeter is shown.

The normal convention for a control knob is clockwise for maximum, and anti-clockwise for minimum, and if the circuit is connected as fig. 8, this should come out correctly. Should you find the reverse happening, break the contact between the voltmeter and 1, (shown as X on the diagram), and couple the voltmeter connections to point 2 on the diagram. This should correct the fault.

Fig. 9 shows the back view of a potentiometer. Tags 1 and 2 are the two ends of the variable resistor, and tag 3 is the sliding arm.

Other methods of providing a DC supply for electroplating are large 30 amp battery chargers for milk floats.

16

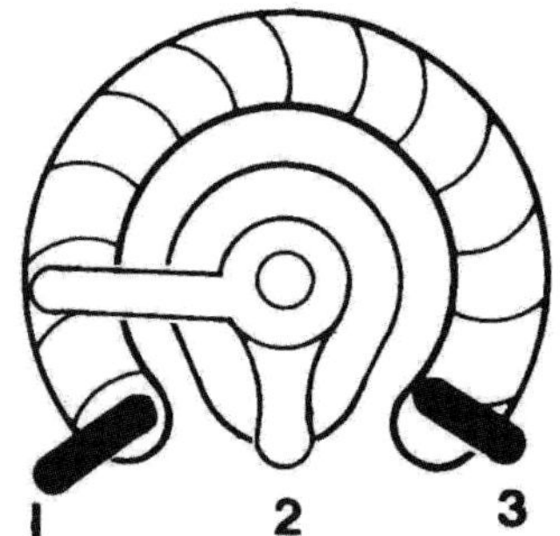

Fig. 9

COMPONENTS LIST

FOR 6 VOLTS	
POTENTIOMETER	25 OHMS, 25 WATTS
VOLTMETER	6 VOLTS
AMMETER	1 AMP F.S.D
SWITCH	5 AMPS

FOR 12 VOLTS	
POTENTIOMETER	50 OHMS 50 WATTS
VOLTMETER	12 VOLTS
AMMETER	1 AMP E.S.D
SWITCH	5 AMPS

(Daveyset types have been used by the author).

The simplest way of supplying a current for electroplating is a 12 or 6 volt battery connected in series with an ammeter and switch. The limitation of this is that the battery has to be re-charged after a period of time, depending of course on the amperage drawn.

CHAPTER 3

The Electroplating Tank

The electroplating tank, vat, or bath, whatever it is called, contains the electrolyte. However, other parts of the process have to be considered, such as cleaning and rinse tanks.

The electroplating and cleaning tanks are the most important, because they hold corrosive solutions of chemicals.

The rinse tanks, as their name suggests, are merely suitable vessels containing plain water.

The main points to consider are the materials, the construction, the size, and if requiring to be heated. This applies to both cleaning and electroplating tanks.

Taking the points in order:-

MATERIALS

Most chemical solutions are either acid or alkaline. The only neutral substances, i.e., p.H 7.0, are usually organic solvents (used in the initial cleaning) and water.

Listing the cleaning solutions and electrolytes used in various chapters with the relevant materials will illustrate what can and cannot be used for a particular solution (see Table 1).

NOTES REGARDING TABLE 1

(1)   Where plastics are mentioned, the most usual are polythene and polypropylene. These plastics come under various trademarks, and can all be used. Good examples of these are half and one gallon ice cream cartons. Other good plastic containers for tanks are cut-down 2.5 liter chemical containers or ten gallon carboys.

(2)   Using mild or stainless steel tanks has the advantage that they can be heated easily with a gas ring. They also have more rigidity, especially when heated. This applies to the alkaline cleaners.

(3)   Pyrex-type glass is quoted because it is heat resistant. When it is in the form of chemistry laboratory squat beakers it can be heated on an electric hotplate or a Bunsen burner. Custom-made plastic tanks can be purchased from electroplating supply houses.

Small plastic tanks can be heated by low wattage aquaria plastic-covered heaters. For larger volume tanks, varying sized wattage of silica sheathed heaters up to 2KW, or stainless steel or titanium immersion heaters of similar wattage, can be used.

Table 1 Recommended Material for Tanks

	Electrolytes	Remarks
N°1 Zinc	Plastic, propylene type, Pyrex glass, stainless steel	Mildly acidic electrolyte
N°2 Zinc	Mild steel, Plastic, propylene type, Pyrex glass, stainless steel	Alkaline
Dull Nickel	Plastic, propylene type, Pyrex glass	
Semi-bright Nickel	Plastic/rubber lined steel, Pyrex	Mildly acidic
Dull Copper	Plastic, propylene type, Pyrex glass, stainless steel.	
Semi-bright Copper	stainless steel, Plastic/rubber lined steel	Acidic
Tin	Mild steel, Plastic, propylene type, Pyrex glass, stainless steel	Alkaline
Aluminum Anodize	Plastic, propylene type, Pyrex glass, Lead-lined	Acidic

Pretreatment Solutions

N°1 Soak Cleaner	All this group mild steel	All Alkaline
N°1 A Electrolytic Cleaner	Plastic, propylene	N°1 A Mild Steel can be used
N°2 Cleaner Aluminum	Pyrex glass, stainless steel	for Anode and Cathode
N°1 Hydrochloric Acid Pickle	All this group	All Highly
N°1A Hydrochloric Acid Pickle	Plastic, propylene type, Pyrex glass	Acidic
N°2 Sulphuric Acid Pickle	Stainless steel	
N°3 Pickle Aluminum	Earthenware (glazed)	
N°4 Bright Dip		
N°5 Bright Dip Aluminum		
Zincate Dip	Plastic, propylene type, Pyrex glass, stainless steel	Highly Alkaline

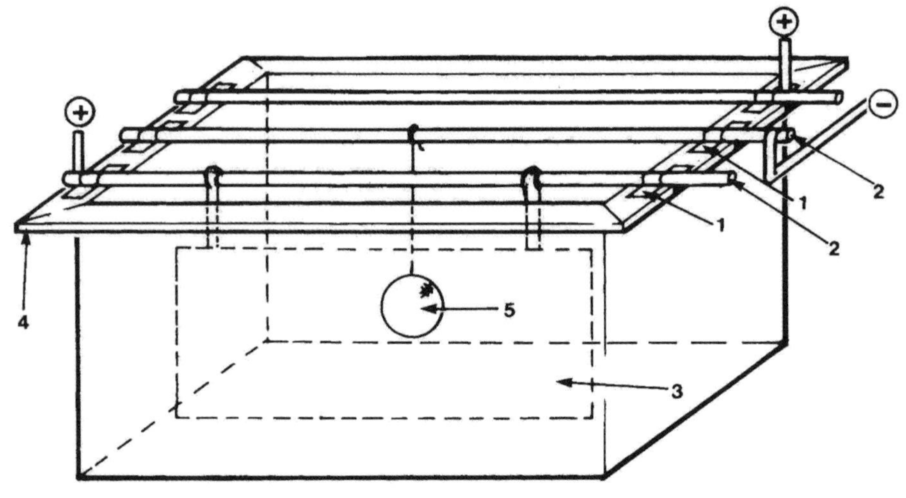

Fig. 10 *Typical electroplating tank.*

TANK CONSTRUCTION

In the construction of electroplating and cleaning tanks for professional finishing shops, glass and most plastic tanks are made by molding, and these are purchased according to the size and volume required. Some plastic can be welded, such as polythene and polypropylene. Be careful to check that these tanks have no leaks. It is good practice to fill them with water and allow to stand for a day, to see if any leaks appear.

For metal tanks that have been welded the same checking for leaks applies.

For stainless steel tanks make sure a good grade of stainless steel is used, such as 317S12. This will stand the corrosive nature of the bright dips and pickles.

A typical tank is illustrated and the following numbers relate to that shown in Fig. 10:

(1) Insulated holders for anode and cathode bars, usually made of porcelain or plastic. If the tank is made of steel, these holders then keep the anode and cathode bars insulated. These are bolted or screwed onto the flanges. For plastic, the tank material provides insulation.

(2) The anode and cathode bars, usually made of copper or brass, typical diameters being ¼in. ½in. to 1in. and 1½in., depending on the size of the tank and the weight of the anode they will have to support.

(3) Anode. These can be sheet, drilled and hooked, or hooked anodes. Make sure the hooks are out of the electrolyte.

(4) The rim, or the flange around the top of the tank, is useful for handling and affixing the holders for the anode and cathode bars, and

20

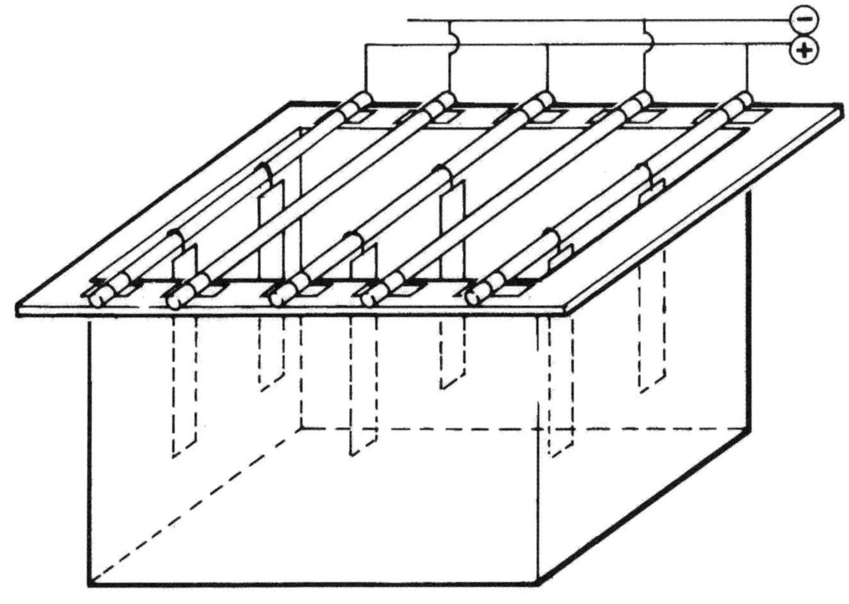

Fig. 11 *Alternative arrangement of anode and cathode bars.*

holders for the various heaters and agitation equipment.

(5)   This is the cathode, or component to be electroplated. It can either be wired on with copper wire or aluminum wire for plating or anodizing. The component can be jigged or hooked. A useful method for small components is to pre-clean and spread them out on a piece of stainless steel mesh shaped like a basket, with a wire or hook through the middle (Fig. 12). After spreading them over the surface of the mesh, the basket is immersed in the electrolyte suspended by the hook from the cathode bar. Shake the basket at various times to move the components. This will prevent areas being unplated.

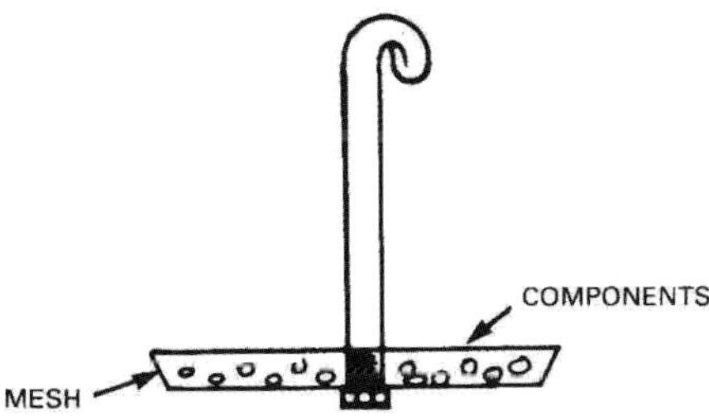

Fig. 12 *Stainless steel basket for small components.*

The calculation of the volume of a tank is:-

Volume (V) in gallons with the dimensions in inches.

$$V = \frac{\text{length x width x height}}{276.5} \text{ cu – inches (inches).}$$

$$V = \frac{\text{length x width x height}}{0.16} \text{ cu – feet (feet).}$$

$$V \text{ in liters} = \frac{\text{length x width x height}}{1000} \text{ cu-centimeters (cc).}$$

These dimensions will be the total volume to the measured height of the tank. The actual volume is to the height of the solution.

If agitation of the electrolytes is needed (this depending on what type of finish is required) two methods are suggested:-

(1)   For nickel and copper electrolytes, as well as the anodizing, they can be agitated by an electric paddle stirrer situated at the side of the tank.

(2)   A good method is to place a plastic pipe at the bottom of the tank, with small holes drilled in. Attached to this, by means of a flexible pipe, is a regulated compressed air source. When air is passed through the pipes, this gives a bubbling action, which agitates the solution. Care – regulate well, or else it will froth and bubble over.

Housekeeping with electroplating process tanks

(1)   Always remove the anode and cathode bars, and clean down by rubbing with scouring powder, Scotchbrite or cloth. Rinse in water and replace. Check all electrical contacts on the tank for heat. If hot, make sure that a better connection is obtained. (Heat dissipates the current).

(2)   Make sure all the solutions are up to the required volume. If not they are topped up with clean water, or distilled or deionized water.

(3)   Make sure all the anodes are clean, and the contacts to the anode bar are clean. In all these areas the current density will vary if there is a bad contact.

(4)   Ensure all rectifier electrical contacts are clean.

(5)   Remove the anodes and rinse when the electrolyte is not in use.

(6)   Cover the electrolytes with lids when not in use.

CHAPTER 4

The Cleaning of the Substrate

The most important part of the electro-plating process is the preparation and cleaning of the substrate, i.e. the surface of the component prior to the electro-plating operation.

This part of the process determines the appearance and the adhesion of the electroplated deposit, as well as its anti-corrosion properties.

The important point to remember about a cleaning cycle is to make it as simple as possible within the confines of the component to be cleaned, the required appearance of the electroplated deposit, and what metal is to be electrodeposited. Elaboration of the cleaning process can lead to a poorly finished component, and consequently lead to a waste of materials.

The first part of any cleaning cycle is de-greasing. This part of the cleaning operation is carried out with an organic solvent and referred to as the solvent clean. There are various de-greasing agents used in this part of the cycle and they are usually some form of paraffin, white spirit, industrial alcohol, or halogenated hydrocarbons such as Genklene.

In industrial finishing shops a vapor clean is used in custom-built tanks containing halogenated hydrocarbons such as trichloroethylene. These are used within certain health and safety regulations, which makes them an expensive capital process.

For general use, to remove soil, grease and oil etc. (all being soluble in organic solvents) the components are wired, hooked or placed in a metal basket, usually stainless steel, which is then immersed in a metal container containing the solvent and agitated from one to ten minutes, depending on the amount of grease and soil. They are then shaken and allowed to air dry. Large components can be wiped with a solvent-soaked cloth before dipping. Great care must be taken with solvents as they are flammable, give off dangerous vapors and remove grease from the skin and could cause dermatitis. It is best to carry out this operation in an open atmosphere, with no naked flames and using gloves.

The solvent clean is a preliminary clean which is carried out on all metals.

ALKALINE CLEANERS

Alkaline Soak Cleaner

This cleaning solution can be used for all ferrous metals, copper and its alloys. It removes the last traces of grease and oil and residual polish compounds from the components.

No.1 Alkaline Cleaner

Make up:-

Sodium Hydroxide 6oz. per gallon 40 grms. per liter.

Sodium Carbonate 4oz. per gallon 25 grms. per liter.

Sodium Trisilicate 4oz.per gallon 25 grms. per liter.

This can be made up with water to the required volume. Care must be taken with sodium hydroxide which is very alkaline: gloves and goggles must be used.

The cleaner is used at a temperature of between 60°C and 80°C, (140°F to 176°F). For ferrous metals it is best to use the temperature of 80°C. It is recommended to use a steel container for the cleaner. The components are immersed from two to ten minutes.

For a *Second Stage Cleaner* or *Electrolytic Cleaner* the above bath can be used. It can be used in the same tank when made of steel. It is considered good practice to place two tanks in tandem, the first for a soak cleaner, and the second an electrolytic cleaner.

For ferrous metals the tank is made anodic, that is the tank is connected to the positive part of the electrical supply and the wired or hooked components connected to the cathodic negative part of the electrical supply. Make sure that they are insulated, by taping across the top of the tank.

The high end of the temperature range is used, that is 80°C at a current density of 10 a.s.f. – 30 a.s.f. (1.0a/dm^2 – 3.0a/dm^2) from one to two minutes.

The mechanism of this type of cleaning is that hydrogen gas is liberated over the surface of the cathode, i.e. the component, the surface of which is "scrubbed" by the hydrogen gas and this in consequence effects a cleaning action.

For copper and its alloys the same procedure is used as for the ferrous metals, but the temperature used is 60°C (140°F), the lower end of the temperature range. The current density is also lower at 5 a.s.f. (O.5a/dm^2), for a period of twenty to forty seconds.

To finish off this part of the cycle for both ferrous materials and copper and copper alloys the polarity is reversed, i.e. the component made anodic and cleaned for a further ten seconds at the same current densities as for ferrous and copper and copper alloys respectively. This removes a tiny amount of metal, hence giving a clean and active surface.

Alkaline Cleaner for Aluminum and Zinc Die Castings

These are cleaned in a low alkali cleaner, which gives a dull, frosty appearance to the components but offers a good clean surface prior to electroplating or anodizing.

No. 2 Alkaline Cleaner

Make up:-

Sodium hydroxide 3oz. per gallon 20grms. per liter.

Sodium Carbonate 4oz. per gallon 25grms. per liter.

The cleaner is used at a temperature of 60°C (140°F). However, it can be used at a lower temperature if a reduction in the frosty appearance is desired. The immersion time is from one minute to three minutes, depending again on the appearance required.

The tank for this cleaner can be of plastic, steel or glass.

PICKLES

The next part of the cleaning cycle is the pickling process. Pickling metals means the removal of impurities such as scale, and, in the case of steel, rust, from the surface, with little or no attack or removal of the actual metal underneath the impurities.

Pickles are formulated from mineral acids such as hydrochloric and sulphuric acids.

No. 1 Pickle

Hydrochloric Acid Pickle.
Make up:-
Concentrated Hydrochloric Acid 31.5fl. oz. 200 mls.
Water made up to one gallon (or 200ml in one liter).

This pickle is used at room temperature. The recommended tank to use is of plastic, usually polypropylene, or glass. If a steel tank is used, it must be lined with rubber or plastic. It is preferable to have a lid on when not in use because of the fumes, which will cause rusting of steel objects in close proximity.

The procedure for making up is to half fill the tank with cold water, then pour in the measured volume of concentrated hydrochloric acid *slowly*. Take care to use gloves, goggles and overalls and carry out in an open space. After the acid has been added, make up to the required volume. The reaction of hydrochloric acid with water does not raise the temperature.

This pickle can be used both for ferrous materials and copper and its alloys.

The immersion time varies according to how much scale is present on the components. For normal scale removal, one to four minutes is the usual time. The components are usually wired up with copper wire, hooked, or on jigs.

To make up an *Inhibited Pickle,* the make up for the hydrochloric pickle is used with 0.5% by weight Hexamine addition.

No. 1(A) Pickle

O.8oz. per gallon or 5grms. per liter of Hexamine.

The result of adding the Hexamine will be the removal of the scale by the pickle, but there will be no action of the pickle on the clean metal surface. In other words, the action of the pickle will cease when the scale has been removed. NOTE – this inhibited pickle can only be used on ferrous materials.

No. 2 Pickle

Sulphuric Acid Pickle.
Make up:-
Concentrated Sulphuric Acid 9fl.oz. 55 mls.
Water made up to one gallon (55ml in one liter).

The same material is used for the tanks as in No.1 Pickle. The procedure for making up the pickle is also the same as No. 1 Pickle, but take care with sulphuric acid – use goggles, gloves and overalls, as this is corrosive, and when added to water the temperature of the solution rises rapidly, so stir continuously while slowly adding the acid. (See the difference with No.1 Pickle where no heat is generated).

After cooling, the pickle is ready for use. It is used at room temperature and used both for iron and steel and copper and brass.

For normal scale removal, one to four minutes immersion time is required.

No. 3 Pickle

Aluminum and Alloys Pickle.
Make up:-
Concentrated Nitric Acid 39fl.oz. 250mls.

Sodium Fluoride 1.5oz. 10grms.
Water made up to one gallon (250ml in one liter).

This is used as a pickle for aluminum, or for de-smutting certain aluminum alloys, and generally imparts a clean surface.

The bath can be made up in a glass or plastic tank, i.e. polypropylene.

To make up the pickle, half fill the tank with water, add the measured amount of concentrated nitric acid to the water, slowly, continuously stirring. Next, add the sodium fluoride and stir well until dissolved. Finally, make up to the required volume with water. CARE with nitric acid, being corrosive; use gloves, goggles and overalls.

The pickle is used at room temperature, with immersion time of between thirty seconds and one minute.

Bright Dips
No. 4 Dip.
Make up:-
Concentrated Sulphuric Acid 80fl.oz. 500mls.
Concentrated Nitric Acid 30fl.oz. 185mls.
Concentrated Hydrochloric Acid 0.5fl.oz. 15mls.
Water 50fl.oz. 300mls.
(Making one gallon or one liter respectively).

The tank can be made of glass, plastic, polythene, P.V.C., or a good quality stainless steel such as 317S12 grade.

To make up the bright dip, the water is poured in the tank, and the concentrated sulphuric acid added slowly, stirring continuously; watch for overheating. The solution is allowed to cool to room temperature. The measured amount of concentrated nitric acid is added, stirring continuously, then the measured amount of concentrated hydrochloric acid added.

It is good practice to stand plastic containers in an outer tank containing cold water.

Care, when making up this bath. The concentrated mineral acids are corrosive and cause burns, therefore gloves, goggles and overalls must be worn.

The bright dip is used at room temperature for copper and copper alloys and nickel silver only. The immersion time is a matter of seconds, this being dependent on the surface required. After bright dipping, they must be immersed in a cold water rinse to remove the acid. Put a small amount of sodium carbonate in the rinse to neutralize, say 2oz. per gallon. The components to be bright dipped are wired with copper wire. Small components can be dipped in stainless steel baskets. This process must be well ventilated, or done in the open air, because when dipping the components, red fumes of nitrogen dioxide are given off from the dip, which are extremely hazardous.

For Bright Dipping of Aluminum
No. 5 Dip.
Make up:-
Concentrated Phosphoric Acid 15.4fl.oz. 440mls.
Hydrogen Peroxide (20 volumes) 0.14fl. oz. 4mls.
Water 0.73fl.oz. 21 mls.
The solution is operated at a temperature of 90°C(195°F).

The tank used is made of glass or plastic, i.e. propylene. The water is poured in the tank, and concentrated phosphoric acid added slowly, stirring continuously. After mixing, the solution is allowed to cool, then the hydrogen peroxide is added. Care. Concentrated phosphoric acid and hydrogen peroxide are corrosive, therefore gloves, goggles and overalls must be worn. The aluminum components are immersed from ten seconds to one minute, depending on the brightness required and the particular aluminum alloy being dipped. Immediately after dipping the components are rinsed in cold water to remove traces of the dip.

Outlined in this chapter are a number of chemical cleaning methods. These give a chemically clean surface, prior to electroplating, which is the ideal situation.

However, various physical methods can be used with or in some cases instead of chemical cleaning.

Wet scouring powder, Scotchbrite pads and wire wool can be used on components, especially copper or brass ones. These can then be rinsed and directly electroplated, or put through the various chemical cleaning processes.

Grit or wet blasting is an excellent method of cleaning components, especially cast iron, cast components or heavily rusted steel articles. Components cleaned this way can be directly electroplated. In the case of cast components, this reduces the risk of occluded cleaners leaching out after electroplating.

After the solvent clean process, if desired, certain areas which may not need to be electroplated may be 'masked off'. This is carried out by using masking tape, PVC, or similar plastic tape, or 'stopping off' lacquers. These are lacquers made of synthetic resins, such as polyurethane varnish. They are painted on and allowed to dry. It is best to use a tape or lacquer commensurate with the hottest part of the electroplating process or the masking may break down during processing.

POINTS TO REMEMBER

Always add acid to water, not vice versa. After mixing with water it is usually less hazardous.
Always wear gloves, goggles and overalls.
If there is any spillage on parts of the body, wash with running cold water.
Work in the open air, or have good ventilation.

TABLES

The accompanying tables cover the cleaning and the electrolytes used in the deposition of the metals. It is a summary of chapters four and five, and parts of chapters six and nine.
It shows the substrate materials with the combination of the cleaning cycles, the electrolytes, and the options to achieve which finish you require.

Key to reference numbers on the tables.

(1)   The preferred clean is a grit, or wet blast, prior to electroplating.

(2)   To electroplate either of the acid coppers a minimum of a flash of either of the nickels must be deposited.

(3)   For tin electroplating, to facilitate soldering, it is recommended that a flash of nickel be deposited before tin electroplating. This is particularly necessary on brass, because it prevents de-zincing of brass after the solder operation.

(d)   The zincate dip is necessary when electroplating aluminum with nickel and other metal deposits electrolytically.

SUBSTRATE MATERIAL	Solvent Clean	N°1 Soak Cleaner	N°1A Electrolytic Cleaner	N°2 Alkaline Cleaner Aluminum	N°1 Hydrochloric Acid Pickle	N°1A Hydrochloric Acid +Hexamine	N°2 Sulphuric Acid Pickle	N°3 Pickle Aluminum
IRON	Yes	Yes	Yes	–	Yes or 1A or 2	Yes or 1 or 2	Yes or 1 or 1A	–
STEEL	Yes	Yes	Yes	–	Yes or 1A or 2	Yes or 1 or 2	Yes or 1 or 1A	–
HIGH TENSILE STEEL	Yes	Yes	Optional	–	Yes or 1A, 2	Yes or 1,2	Yes 1,1A	–
IRON/STEEL[1] CASTINGS	Yes	Yes	Yes	–	Yes or 1A, 2	Yes or 1,2	Yes or 1, 1A	–
ALUMINUM & ALLOYS	Yes	–	–	Yes	–	–	–	Yes
ZINC DIECASTING	Yes	–	–	Yes	–	–	–	Yes
BRASS	Yes	Yes	Optional	–	Yes or 2	–	Yes or 1	–
COPPER	Yes	Yes	Optional	–	Yes or 2	–	Yes or 1	–
BRONZE	Yes	Yes	Optional	–	Yes or 2	–	Yes or 1	–
NICKEL SILVER	Yes	Yes	Optional	–	Yes or 2	–	Yes or 1	–

CONTINUATION

SUBSTRATE MATERIAL	N°4 Bright Dip	N°5 Bright Dip Aluminum	Zincate Dip	N°1 & 2 Zinc	Dull Nickel Semi Bright Nickel	Dull Copper Semi Bright Copper	Tin	Anodize
IRON	–	–	–	Yes	Yes	Yes[2]	Yes[3]	
STEEL	–	–	–	Yes	Yes	Yes[2]	Yes[3]	
HIGH TENSILE STEEL	–	–	–	Yes	Yes	Yes[2]	Yes[3]	
IRON/STEEL CASTINGS	–	–	–	Yes	Yes	Yes[2]	Yes[3]	
ALUMINUM & ALLOYS	–	Optional	Yes[4]	Yes[4]	Yes[4]	Yes[4]	Yes[4]	Yes
ZINC DIECASTING	–	Yes	–	–	–	Yes	–	–
BRASS	Optional	–	–	Yes	Yes	Yes	Yes	–
COPPER	Optional	–	–	Yes	Yes	Yes	Yes	–
BRONZE	Optional	–	–	Yes	Yes	Yes	Yes	–
NICKEL SILVER	Optional	–	–	Yes	Yes	Yes	Yes	–

CHAPTER 5

The Electrolyte

The term electrolyte is defined as the conducting medium for most electroplating processes. The most common electrolyte is an aqueous solution with water soluble chemicals. When a current is allowed to flow through the solution via the anode and cathode, a deposition of metal occurs at the cathode. The electrolytes vary in p.H and may be acidic solutions, neutral and alkaline solutions.

The electrolyte must contain the dissolved salt of the metal to be deposited. The salts dissolve in water and form ions. For example, copper sulphate dissolved in water forms Cu^{++} SO^{--} , the ions of Cu^{++} and SO_4^{--}.

Generally, the more complex the ion, the more efficient, and a much smoother deposit of the metal occurs. Most professional electrolytes are of the cyanide ion, which is complex, as distinct from the simple ion of copper sulphate. The main reason why cyanide electrolytes are commonly used is that with a complex ion, the actual content of the metal in the ion is relatively low compared to a simple ion, and this retards the formation of an immersion deposit when the cathode is placed in the electrolyte, which can cause problems with adhesion.

A good example of this is the dipping of a piece of steel in acidic copper sulphate, which results in an immersion deposit of copper.

For a complex ion of copper cyanide the % of copper present in the solution is:- $[Cu(CN)_4]^{---}$ molecular weight = 167.5. Atomic weight of copper is 63.5.

% copper $\dfrac{63.5 \times 100}{167.5}$ = 38% copper cyanide in electrolyte 15 gms/liter

Gives 5.7gms of copper.

For a simple ion of copper sulphate the % of copper present in the solution is:- $CuSO_4$ molecular weight = 159.5.

% copper $\dfrac{63.5 \times 100}{159.5}$ = 40% copper sulphate electrolyte 200gms/liter

Gives 80gms of copper.

Coming to more practical terms for electrolytes, the most common are the ones used for zinc plating, in their various forms. The one that I have used, and is considered reasonably safe for use in a workshop or garage, is a zinc chloride bath.

ZINC CHLORIDE BATH

This is a simple electrolyte to use and maintain and has the advantage of electroplating on difficult metals, such as

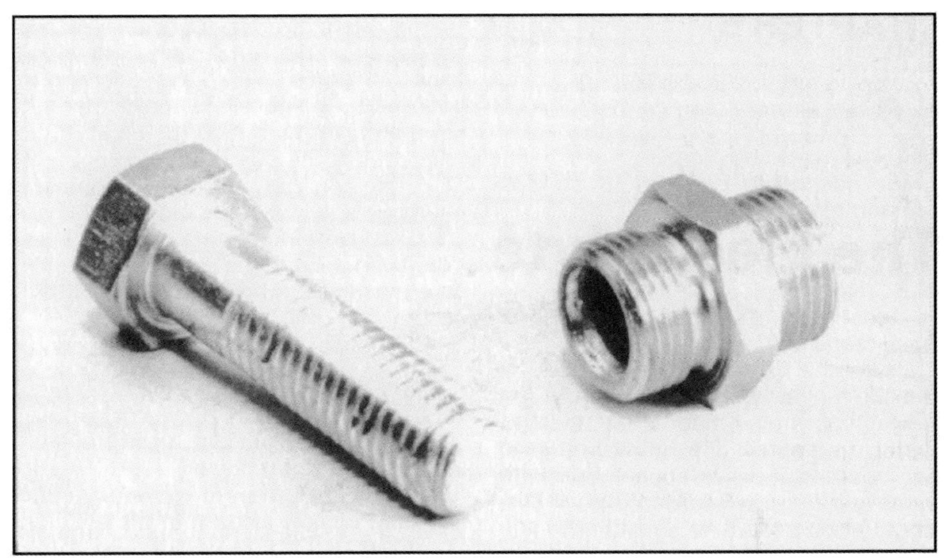

Examples of zinc and passivate and semi-bright nickel. The bolt is electroplated with zinc, and blue passivated. The adaptor is electroplated with semi-bright nickel from the electrolytes described in this chapter.

cast or malleable iron. It is operated at room temperature, therefore no heating costs are incurred, and the deposits are easily treated with certain chemicals to give a colored finish (passivate).

The formulation of the electrolyte is:-

Zinc Chloride 3oz per gallon 20grms per liter.

Ammonium Chloride 20oz per gallon 120grms per liter.

To make up one gallon of electrolyte add six pints of tap water to the plating tank, then add the 20oz of ammonium chloride, stirring well until completely dissolved.

In another container, put two pints of warm tap water, and add 3oz. of zinc chloride, stirring until dissolved. Pour the two pints into the six pints already in the plating tank and mix well.

The solution is now ready for use. It is advisable to mark the one gallon level, or whatever volume you use, on the outside of the tank with a waterproof marker pen, so that as the level of the tank falls through evaporation or drag out it can be topped up again.

If you can keep a lid on the tank when not in use this will minimize evaporating and prevent dust falling into the solution.

Operating Conditions
The tank is connected up to the power supply, the positive connected to the anodes, in this case pieces of zinc 4in. x 3in., or perforated zinc sheet 6in. x 6in. bought from your local ironmongers. Pre-clean the components, and immerse in the electrolyte. Turn the current on, and regulate to the current you require

within the current density range for the electrolyte.

For this electrolyte the plating current should be between

14.4 a.s.f. to 28.8 a.s.f.

or l.54 a/dm^2 to 3.1a/dm^2.

The electrolyte is operated at room temperature 15°/20°C.

At 14.4 a.s.f. the zinc deposited will be 0.001 in. or 25 microns, at a plating time of 83 minutes.

At 28.8 a.s.f. a deposit of 0.001 in. or 25 microns is 42 minutes.

Using the lower end of the current density range for electroplating will give a finer, more even deposit of zinc. The faster the deposition, the more uneven and coarse the deposit becomes. Occasionally it may be necessary to electroplate a component of small area, such as a small washer or pin. The current output may be too large at the lowest range of the supply. In this case, hang a piece of steel plate or two pieces of plate (robbers) each side of the component. This will increase the surface area to correspond with the electroplating current, and thus bring it into the current density range. The robbers will also carry out a second function by removing the high current density, giving a more even and smooth appearance, i.e. prevent "burning".

The p.H of the electroplating solution
The optimum is between p.H 3 to p.H 4, but the solution performs quite well up to p.H 7. Strip papers can be used to check the p.H. These can be purchased from chemical supply houses and large electroplating equipment suppliers. Always keep the test papers well sealed when not in use. To use the papers you will see that they are numbered 1 – 14, universal type, and that between the numbers 6 and 7 is a strip. Immerse the test paper in the electrolyte for a couple of seconds, and note that the strip between 6 and 7 changes color. Match this color to one of the numbered colors either side of the center strip, and the number that matches is the p.H of the solution. The p.H can also be measured by one of a number of p.H meters and portable p.H 'sticks' available on the market. These vary in price, and the portable p.H sticks are reasonably priced between £40 and £130.

To obtain a satisfactory deposit in most electrolytes it is necessary to 'electroplate the electrolyte in'. This is done by hanging a piece of scrap steel in the solution, and, using a current within the current density range, leave it electroplating for about one hour. This electrolyzes the solution, and takes out some of the impurities. However, to take out impurities as a specific operation, the electrolyte is plated out at a very low current. This induces the impurities to deposit out first on the scrap piece, then the normal metal of the electrolyte is deposited.

ZINC HYDROXIDE BATH
This is an alkaline zinc electrolyte made up with sodium hydroxide. This chemical is extremely caustic, and care must be exercised when using it. Goggles, gloves and overalls must be worn when handling, and also keep away from children and animals. When not in use it is kept in a tightly closed plastic or glass jar, because it takes in water from the atmosphere and decomposes to sodium carbonate and water.

The formulation of the electrolyte is:-
Zinc Oxide 1.0oz per gallon 6grms per liter
Sodium Hydroxide 9.0oz per gallon 55grms per liter
Dextrin 1% addition by weight.

To make up the solution add the weighed-out amount of sodium hydroxide (9.0oz.) to six pints of water and stir until dissolved. Next, add the weighed-out zinc oxide (1.0oz.) and stir until dissolved. The last to be added is the dextrin. This is stirred in until dissolved. The volume is then made up to one gallon with water, and the electrolyte is ready for use, but, as mentioned in the previous zinc electrolyte, it must be 'plated out' before it is usable. For this solution the 'plating out' requirements are 0.2amps for 12 hours with a piece of steel 2in. x 2in.

Operating Conditions

The same conditions apply as for the previous zinc chloride electrolyte. The electroplating current density range is between
2 and 20 a.s.f. or 0.2a/dm^2 to 2.0a/dm^2.

The electrolyte is operated at room temperature 15° – 20°C.

Being alkaline p.H 12 – 14, the control of the p.H is not needed with this solution.

GENERAL FAULTS IN ZINC ELECTRO-PLATING

FAULT The deposit is rough or coarse in texture, and may be discolored.
REASON Current density too high.
REMEDY Reduce the current density.

FAULT Rough deposits can also be caused by suspended matter in the electrolyte.
REMEDY Filter the electrolyte through a filter paper or fine cloth.

FAULT Deposits rough, and electroplating sluggish (i.e. lack of thickness).
REASON Low conductivity of electrolyte.
REMEDY To No.1 zinc electrolyte add

ammonium chloride to the solution at the rate 2 to 4oz. per gallon. For No.2 zinc electrolyte 0.25oz. of zinc oxide and 1 to 1.5oz. of sodium hydroxide per gallon of solution.

FAULT Electrolyte appears to be a rusty color..
REASON Iron from the components being electroplated is dissolved into the solution.
REMEDY For No.1 zinc electrolyte add 50 milliliters of hydrogen peroxide, stir well, and leave to settle. When settled, carefully decant off the clear solution.

FAULT The deposit is patchy.
REASON The pre-treatment clean is faulty.
REMEDY Strip the deposit off to the metal by immersing the component in 30% sulphuric acid or 15% hydrochloric acid until all the zinc is removed. Go back through the pre-treatment, and replate.
CARE A small amount of hydrogen gas is given off, so make sure the area is well ventilated. Using sulphuric or hydrochloric acid, goggles and gloves must be worn.

COLORING ZINC ELECTROPLATING

Coloring zinc electroplating is usually called passivating the zinc surface. As well as being decorative with the various colors, it enhances the performance of the electrolyte by increasing its anti-corrosive properties.

Zinc, along with cadmium, are classed as 'sacrificial coatings' on ferrous materials. This means that the electroplated deposit will be oxidized to atmosphere instead of the component rusting or forming iron oxide, but over a period of years the electroplated deposit will disappear from the component. To reduce the active deposit, a 'passive' layer, usually of chromium ions, is applied. This reduces the activity, hence prolonging the life of the deposit and enhancing the anti-corrosive properties.

For the different colorings the baths are made up as follows:-

Black

Make up *Ammonium Molybdate* 4oz. per gallon. 25grms per liter.
Concentrated Ammonia (.880 S.G.) 6 fluid oz. per gallon. 37.5mls. per liter.
Water to make up to a gallon (or liter).

Leave the component in the solution until the desired shade is obtained. For deep blacks, heat the solution, but note the ammonia will fume and give off a strong smell, therefore it must be well ventilated or done outdoors.

In both cases, when the correct color is obtained, rinse in cold water, then in hot water, and leave to harden off the layer. Do not touch the colored surface until hardened.

Brown

Make up *Double Nickel Salts* 0.5oz. per gallon 3grms per liter.
Copper Sulphate 0.5oz. per gallon 3grms per liter.
Potassium Chlorate 0.5oz. per gallon 3grms per liter.

Use the solution at 60°C. with an intermediate wet scratch-brushing to even out the color. Waxing off improves the color. *Note* Potassium Chlorate is a powerful oxidizing agent, and the powder must be kept in a well stoppered container.

Blue Color (Passivate)

Make up *Sodium Dichromate* 0.8oz. per gallon 5grms per liter.
Concentrated Nitric Acid 3.2 fluid oz. per gallon 20mls. per liter.

Dissolve the sodium dichromate in one liter of water, then add the nitric acid carefully, stirring well in slowly. The bath is operated at room temperature. The immersion time 10 to 20 seconds.

Iridescent Color (Passivate)

Make up *Sodium Dichromate* 10oz. per gallon 60grms per liter.
Concentrated Sulphuric Acid 0.5fluid oz. per gallon 3.0mls. per liter.

Dissolve the sodium dichromate in the volume required, then add the concentrated sulphuric acid slowly and carefully, stirring continuously. The bath is operated at room temperature. The immersion time is 8 to 10 seconds. *CARE* With nitric acid and sulphuric acid, gloves, goggles and overalls must be worn.

The last two formulations are pure passivate coatings, and are left for 24 hours to harden off before further use. After hardening off they can be painted if required. This combination of zinc, passivate and paint offers good corrosion resistance for ferrous metals.

NICKEL ELECTROLYTES

Nickel offers good corrosion resistance when electroplated on both ferrous and non-ferrous metals such as copper and alloys of copper.

The mechanism of nickel, copper and tin electroplating is that the deposits are more 'noble' than steel, and steel becomes anodic and therefore dissolves. Thus it is important to have good precleaning and to avoid pores and discontinuities on the electroplated surface.

This is distinct from the mechanism of the zinc electroplate, which is less noble than steel, and in consequence the steel

becomes cathodic with the zinc dissolving i.e. sacrificial.
A good all round nickel electrolyte is as follows:-

Watt's Nickel

Nickel Sulphate 38.5oz.per gallon 240grms per liter.
Nickel Chloride 7.2oz.per gallon 45grms per liter.
Boric Acid 4.8oz.per gallon 30grms per liter.

To make up the electrolyte, warm up three quarters of the volume of water, add the weighed amount of nickel sulphate, stir, and warm until dissolved. Then add the weighed amount of nickel chloride, repeat until dissolved. Finally, add the boric acid, which will take time to dissolve. After all the chemicals are in solution, make up to the required volume with water.

The p.H should be checked by whatever means, papers or p.H meter. The p.H should be 3.5.

It is recommended that this solution be 'plated out' at a low current with some scrap steel plate similar to the No.2 zinc electrolyte. This will remove the dissolved impurities in the nickel salts.

The anodes used in this electrolyte are 4in. x 4in. slabs of pure nickel. These can be obtained from electroplating supply houses. This size approximates to the anodes used on Hull Cell Apparatus. However, for a large set-up basketed nickel shot or hooked anodes may be used, but these can be expensive.

The operating temperature for this electrolyte is 50°C (130°F).

The heating may be accomplished by various methods. If a stainless steel container is used, a gas ring is placed underneath. If the container is Pyrex glass or plastic, then an aquarium type electric heater is used.

The current density range is from 10 a.s.f. to 40 a.s.f. or 1.0a/dm^{2}to4.0 a/dm^2.

This is as previously mentioned. The part of the range chosen is dependent on the shape of the component and the texture of the deposit required. The lower the current density the finer the deposit.

This electrolyte will deposit:-
0.0001 in.(2.5microns) at 10a.s.f. in 15 minutes or 0/0001in.(2.5microns) at 30a.s.f. in 5 minutes.

This electrolyte gives a good dull soft deposit which will polish up to a high luster. Its properties are not greatly affected by a wide change in p.H, temperature and balance of chemical composition in the bath.

When this deposit is over-electroplated with decorative chrome it gives a pleasing dull chrome effect, similar to the finish on micrometers and similar tools.

Semi-Bright Nickel Electrolyte

This is based on a Watt's Formulation with the addition of an organic brightener.
Nickel Sulphate 38.5oz.per gallon 240grms per liter.
Nickel Chloride 7.2oz.per gallon 45grms per liter.
Boric Acid 4.8oz.per gallon 30grms per liter.
Saccharine 0.32oz.per gallon 2.0grms per liter.

The electrolyte is made up as for the previous nickel electrolyte (Watts), but with the addition at the end of the weighed amount of saccharine. This is constantly stirred when warm, until dissolved. This dissolution may take some time because saccharine is not very soluble.

The same current density ranges apply for this as for the previous electrolyte, as well as the temperature range.

This electrolyte will deposit:-
0.0001in. (2.5 microns) at 10 a.s.f. in 18 minutes or
0.0001 in. (2.5 microns) at 30 a.s.f. in 8 minutes.

With these electrolytes agitation is recommended either by stirring or compressed air bubbled through the solution.

FAULTS IN NICKEL ELECTROPLATING

FAULT Pitting of deposit.
REASON Acidity of solution too high, nickel content low, boric acid too low.
REMEDY Adjust p.H to between 3 and 5. Adjust p.H with aqueous solution of sodium hydroxide until between the limits. Add 3oz. per gallon of nickel sulphate. Add 0.5oz. per gallon of boric acid.

FAULT Not enough coverage of the component.
REASON Electrolyte temperature too low, or low current density.
REMEDY Increase electrolyte temperature to 50°C–55°C. Increase current density.

FAULT Poor adhesion of the nickel to the component, and may be of burnt appearance.
REASON Poor pre-cleaning of the component. Too high p.H (alkalinity). Too high current density.
REMEDY Strip off the nickel plate, depending whether ferrous or non-ferrous. Go through the pre-clean and re-plate. To reduce the p.H to between 3 and 5, add diluted hydrochloric or sulphuric acid, until the correct p.H range is obtained. In the correction of acidity or alkalinity the reagents are added sparingly, and constantly stirred and checked. CARE – With aqueous sodium hydroxide and hydrochloric and sulphuric acids, gloves and goggles MUST be worn.

Nickel deposits from both electrolytes can be buffed or polished to a good luster. It must be remembered that an allowance on the thickness must be made for polishing.

COPPER ELECTROLYTE

The most used electrolytes of copper are the cyanide copper and the acid copper.

The main distinction between the two electrolytes is that the cyanide copper can be used to deposit copper on both ferrous and non-ferrous metals. The acid copper can only be used to deposit copper on non-ferrous metals. For the amateur in the home, the use of cyanide is a considerable hazard, and with the difficulty of obtaining this chemical I have not included any processes in the text. However, a method can be used to finally deposit copper on a ferrous material, which will be described.

A good general copper electrolyte which can be buffed and polished:-
Copper Sulphate crystals 32oz. per gallon 200grms per liter.
Concentrated Sulphuric acid 4.50fl.oz. per gallon 30mls per liter.

To make up the electrolyte warm up three quarters of the volume of water, and add the weighed amount of copper sulphate crystals, and stir constantly until dissolved. To the cooled copper sulphate solution add very slowly dropwise the concentrated sulphuric acid, previously measured. Stir well until mixed well into the solution. Next, top up with water to the required volume, and the electrolyte is ready for use. The current density is 4.a.s.f. to 30 a.s.f.

0.4a/dm^2 to 3.0a/dm^2.
This electrolyte will deposit:-
0.0001 in. (2.5 microns) at 10a.s.f. in 12 minutes.
0.0001 in. (2.5 microns) at 20a.s.f. in 6 minutes.

Agitation is recommended at the top end of the current density range. Pure copper may be used for the anodes. At high current densities there is a risk of the anodes polarizing, with a reduction in the current. The recommended anodes for acid copper electrolytes are the phosphorized type. These are purified copper using a process of removing the impurities of oxygen with phosphoric acid in the anode making stage.

The electrolyte is operated at a temperature of 30°C (86°F). For thin deposits of copper it can be used at room temperature. The higher temperature of 30°C gives a smoother deposit of copper.

Semi-Bright Copper Electrolyte
Make up:-
Copper Sulphate 32oz. per gallon 200grms per liter.
Concentrated Sulphuric Acid 1.4oz. per gallon 27 mls per liter.
Thiourea pinch 0.005grms per liter.
Wetting Agent (Teepol) one drop 1ml per liter.

The electrolyte is made up the same as the dull acid copper, but with lower concentration of sulphuric acid. Before the bath is made up to its final volume, the thiourea and wetting agent are added. After making up to the final volume the electrolyte is ready for use. N.B. the thiourea at 0.005 grms. can be regarded as a 'pinch'.

The electrolyte should be operated at 23°C (74°F) for the best results. At lower temperature the copper deposit is less bright. The electrolyte will deposit:-

0.0001 in. (2.5microns) at 10a.s.f. in 12 minutes. 0.0001 in. (2.5microns) at 20a.s.f. in 7 minutes.

Agitation is recommended either by stirring, or bubbling compressed air through the solution.

The same anodes are used as in the dull acid copper electrolyte.

For depositing copper from these electrolytes on copper and its alloys, and on zinc diecastings, the precleaning treatment as given in the chapter on cleaning is used, and then the copper deposited on the surface.

For ferrous metals, steel etc., a different procedure applies. After pre-cleaning the component, it is electroplated in either of the nickel electrolytes, usually at medium current density, until a 'flash' of nickel is deposited – 0.00005in. The component is removed, quickly rinsed, and while still wet is immersed in either acid copper electrolyte, and copper deposited.

The reason for this is that copper deposited from an acid electrolyte will corrode the steel or iron surface. If nickel is deposited prior to the copper electroplating, no corrosion takes place.

Both these electrolytes will deposit a thick layer of pink colored copper that can be polished or buffed to a high luster.

FAULTS WITH COPPER ELECTROPLATING
FAULT Coarse, burnt appearance of the copper deposit.
REASON Too high current density. Temperature of bath too low.
REMEDY Reduce the current density. Increase the temperature of the solution to that recommended for the electrolyte. This fault may be a combination of the two reasons, so with a small reduction in current density, and increase in tem-

perature, the deposit of copper will be satisfactory.

FAULT Fall in current, and rise in voltage, and a black film formed on the anodes.
REASON Lack of sulphuric acid in the solution.
REMEDY Add dilute sulphuric acid in small amounts until the anode loses its black film.

FAULT Poor coverage of the component.
REASON Poor pre-cleaning or lack of sulphuric acid.
REMEDY Strip the copper, go back through the cycle of pre-cleaning and re-plate. Add a small amount of dilute sulphuric acid. If the fault is caused by a combination of the two reasons, then add the sulphuric acid first.

FAULT Nodular or spiky deposit.
REASON Dirty electrolyte with undissolved particles.
REMEDY Filter through a filter paper or fine cloth.

TIN ELECTROLYTES

Tin deposited from tin electrolytes is used for preventing corrosion on both non-ferrous and ferrous metals, and is also used to facilitate the soldering of components. In some cases it is used as a decorative finish.

The potassium stannate electrolyte produces a deposit of tin that is colored light gray, matte in appearance, but can be buffed to a silvery finish. This is a good all round electrolyte for use as an anti-corrosion finish on steel, or for use on both steel and brass for soldering purposes. The make up is as follows:-

Potassium Stannate Electrolyte

Potassium Stannate 15oz. per gallon 95 grms. per liter.
Potassium Hydroxide 2oz. per gallon 12 grms. per liter.

To make up the electrolyte three quarters of the volume of distilled or deionized water is warmed up in the bath. The weighed amount of potassium stannate is added to the warm water, and stirred continuously until dissolved. Next, the weighed amount of potassium hydroxide is added, and stirred until dissolved. *CARE* – with potassium hydroxide (a strong alkali) gloves, goggles and overalls must be worn.

After the chemicals are dissolved, the electrolyte is made up to the final volume with distilled or de-ionized water. The electrolyte is now ready for use, but it is recommended that some scrap pieces of steel or copper are wired up and connected up to the cathode, and the electrolyte 'plated in' at medium current density for about one to two hours.

The anodes used in this process are pure tin anodes, usually in the form of slabs that can be cut to the required size.

The electrolyte is operated at a temperature between 60°C and 80°C. (140°F and 176°F). No agitation is required.

The electrolyte will deposit:- 0.0001 in. (2.5microns) at 10a.s.f. in 10 minutes. 0.0001 in. (2.5microns) at 20a.s.f. in 6 minutes.

The current density is between 10a.s.f. and 40a.s.f. (1.0a/dm^2 and 4.0a/dm^2).

The p.H of the solution is highly alkaline, and stays alkaline. When operating this solution it must be remembered that it is best that before electroplating, a piece of scrap material (steel), is connected to the cathode bar in the electrolyte. The electrical supply is switched on, and the anodes are placed in the electrolyte, and connected to the anode bar. The scrap piece should start being electroplated immediately. At the same time as the scrap piece is electroplating, wire or hook the components to be electroplated to the cathode bar. This is carried out with the current switched on. This procedure produces an iridescent gold colored film on the anodes, which is part of the mechanism of the electroplating process. Moderate current density is used on the scrap piece.

FAULTS WITH STANNATE TIN ELEC-TROPLATING

FAULT Electrolyte works sluggishly, and the anodes are a gray color.
REASON Initial current density too low to form the correct film. Free potassium hydroxide too low.
REMEDY Remove anodes one at a time, and replace them in the electrolyte. Sometimes a slight increase in the current density is beneficial. Add 0.5oz. per gallon of potassium hydroxide.

FAULT The solution turns a slightly pink color, with a spongy tin deposit.
REASON A build up of potassium stannite in the solution.
REMEDY Add a small amount of hydrogen peroxide dropwise, and stir until the pink color disappears. Strip the tin deposit, pre-clean, and re-plate after adding the hydrogen peroxide.

FAULT Anodes covered in a black film, and current drops off.
REASON The anodes have become polarized, and become covered in tin oxide, usually brought about by too high current density.
REMEDY Reduce current density. Remove anodes one at a time, and scour the anodes with a stiff brush until the film is removed. Replace, until all the anodes have been scoured.

CHAPTER 6

Electroforming and Electroplating on Non-conductors

ELECTROFORMING

Electroforming is a process of electroplating that is used in the manufacture of intricate components that are difficult to fabricate or machine. It is an ideal method for making components to tight tolerances and dimensions.

The thickness of metal deposited is considerably more than conventional electroplating (anything from 0.012in. to 0.080in).

The most common metals deposited in electroforming are copper and nickel.

However, the most essential part of the process, and sometimes the most expensive, is the mandrel, which is the shaped material the metal is deposited on.

Mandrels can be divided into two categories.

a). *Disposable Mandrels.*

These are a type of mandrel that

An example of a disposable mandrel made from low melting point alloy. Not the inserts molded in for affixing Perspex windows, and connecting the cathode bar.

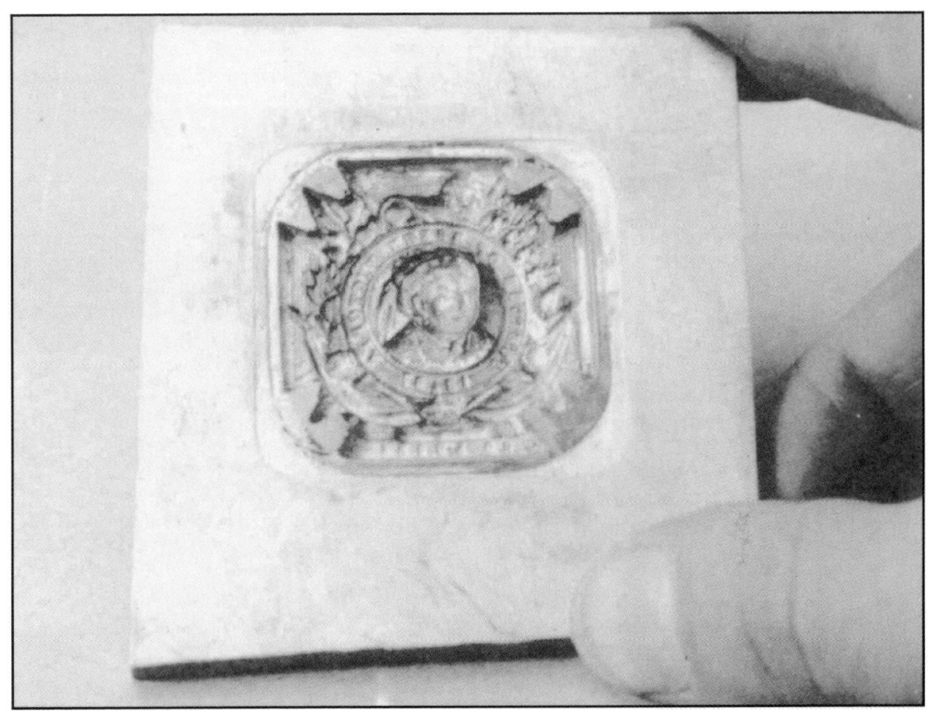

A mold made from a medal by electroforming, set in a backing die allowing it to be used for making a quantity of moldings in various molding materials.

cannot be extracted by pulling out of the electroform, due to being either a bend, or having smaller dimensions at the ends.

The material used can be Perspex, low melting alloys, (Cerrobend, Cerrocast), and some grades of wax. All these materials can be melted or dissolved in solvents. The essential piece of equipment for this mandrel is a mold to cast in. However, with Perspex this is usually machined to shape, and not molded. b). *Permanent Mandrels.*

These are usually made of stainless steel or nickel electroplated steel. Sometimes certain plastics are used.

These mandrels are extractable by pulling on a press.

Other forms of electroforming are quite useful, such as building up worn shafts, or shafts that have been machined down too far.

Small molds can be made for various other processes, by making a replica, electroforming, removing the replica and setting the electroform in a backing die. This mold can be used for making a quantity of components. From a mold like this items can be made and used on, for example, model boats.

The two electrolytes used in electroforming are copper and nickel.

41

The copper electrolyte is the dull copper listed in the chapter on electrolytes.

Copper Electroforming Electrolyte
Make up:-
Copper Sulphate Crystals 32oz.per gallon 200grms per liter.
Concentrated Sulphuric Acid 4.5fl.oz per gallon 30mls per liter.
A small amount of phenol dissolved up in water can be added. This gives grain-refining properties to the electrolyte. The current density is between 10a.s.f. – 20a.s.f. $1.0a/dm^2 - 2.0a/dm^2$.

The operating temperature is 30°C. (86°F).

For nickel, the Dull Watt's electrolyte can be used, but a good electroforming electrolyte is one formulated as follows:-

Nickel Electroforming Electrolyte
Make up:-
Nickel Sulphamate 72oz. per gallon 450grms per liter.
Boric Acid 5oz.per gallon 30 grms per liter.
The same tank can be used as for the dull nickel, and made up the same, with the nickel sulphamate dissolved first, then the boric acid added and dissolved and made up to the final volume.

The p.H is 4.0. To reduce p.H add sulphamic acid. To increase p.H add ammonia 0.880SG. However, being stable, it usually stays at about p.H 4.0.

The solution is operated at 45°C–50°C (113°F – 120°F), with agitation if possible, depending on the thickness required.

The current density is between 10a.s.f. – 50a.s.f. $1.0a/dm^2 - 5.0 a/dm^2$.

Fig. 13

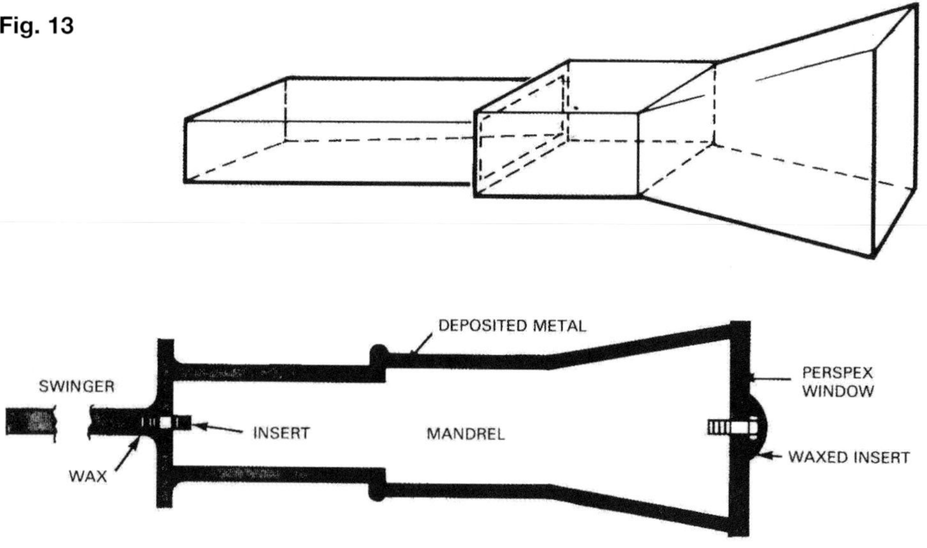

Fig. 14

A simple example of electroforming is a funnel shaped component. See figures 13 and 14.

Figure 14 is the assembly for electroforming the shape. The mandrel is made with inserts (screwed), if molded, from low melting point alloys. If made from Perspex, the holes are drilled and tapped.

Perspex windows are affixed at each end, the hanger attached, and the areas that are conducting are 'stopped off' with molten 'Clamca' wax, or similar material.

If low melting alloys are used, the pre-treatment is a light clean with scouring powder or Scotchbrite, then a dip in the alkaline cleaner and electrolytic cleaner, but the electrolytic cleaner is optional. The mandrel rig is rinsed in water and immersed in the particular electrolyte, and grown to the required thickness.

If Perspex is used, this is made conductive with either a layer of aqueous graphite, or, preferably, silver-loaded conductive paint, which is painted on the surface of the Perspex, *making sure the contact for the hanging bar is well*

An example of an electro-forming mandrel with Perspex window and cathode bar connected and waxed, (top left). Top right is a watch-glass with conducting silver paint. (Note the connecting bar of silver paint on the side). The bottom half of the photograph shows the electroformed shapes, in this case, a bowl and lid electro-formed in silver.

An example of electroplating on aluminum to facilitate soldering. This utilizes zincate solution, followed by a layer of tin, with tinned copper tube soldered onto the final tin layer on the aluminum.

painted. The conductive surface is allowed to dry, and is immersed in the particular electrolyte, and grown to the required thickness.

If nickel-plated steel or stainless steel mandrels are used, i.e. permanent mandrels, the same procedure for cleaning as for low melting point alloy mandrels is followed, and growing is the same.

After growing, the wax and windows are removed, and depending on what the mandrel is made of, it is extracted by the following processes.

Low melting alloy. This is melted, either in an oven, or hot oil until the alloy flows out.

Perspex. The mandrel is immersed in a solvent such as warm Genklene until dissolved up. It may also be extracted by holding the electroform and pulling out the Perspex mandrel.

Nickel-plated steel or *stainless steel.* This is extracted the same as Perspex, pulling from the wide end.

Points to remember on electroforming

(1) If a large quantity of electroforms are required, stainless steel mandrels are more economical.

(2) With permanent mandrels (extractable), these can only be used on electroforms that have shapes that can make them extractable.

(3) A small draft angle should be included on permanent mandrels to aid extraction.

44

ELECTROPLATING NON-CONDUCTORS

Included in this section is the electro-plating of aluminum. This is not strictly a non electrical conductor, but neverthe-less it will not electroplate like conven-tional metals.

After going through the listed pre-cleaning cycle for aluminum, as in chap-ter four, zincate solution is required. This is made up as follows:-

Zincate Solution

Make up:-

Sodium Hydroxide 70oz. per gallon 440 grms. per liter.

Zinc Oxide 13oz. per gallon 87 grms. per liter.

The solution is made up in half the volume of water, adding the sodium hydroxide slowly, and stirring continu-ously. After dissolving the sodium hy-

A molded Epoxy resin figure, sensitized with silver loaded paint, and electroplated.

droxide, and while still warm, add the zinc oxide, stirring until dissolved. Make up to the final volume and allow to cool. The solution is then ready for use.

Ideal containers are Pyrex, glass or plastic. The solution is used at room temperature.

To improve the deposit, that is to make a more even deposit, a small amount of ferric chloride crystals (a large pinch, approximately 0.5grms per liter) and 5 grms of Rochelle salt are added. However, the author has found the original formula satisfactory for most finishes on aluminum.

The immersion time is between ten and thirty seconds. After rinsing in water the aluminum should have a gray appearance.

While still wet from the rinse, immerse in either of the nickel electrolytes to deposit approximately 0.0002in. – 0.0003in., remove, and rinse, and transfer to any other electrolyte you would like as the finish.

This is a useful aid to soldering on aluminum, either all over or selectively, with the aid of masking. The process is as mentioned, but after depositing nickel, the final finish is a deposition of tin 0.0003in. approximately. This finish is ideal for applying solder.

Non-conductors are usually classed as plastics and ceramics, but these can be electroplated when they have been made electrically conductive.

In industry the classic of electroplating non-conductors is the process in the electronics industry of 'plating through hole' on printed circuits. Briefly, this is when the copper clad plastic is drilled. The holes through the middle have plastic faces. This has to be sensitized with various chemicals, such as stannous chloride, then palladium chloride, to make the surface conducive to a layer of copper from an electroless copper solution. After rinsing, the layer of copper is built up in the holes with high throw electrolytic copper, followed by a tin/ lead electroplate.

A simpler method for electroplating on non-conductors was mentioned in the electroforming section, namely aqueous graphite and silver-loaded paint.

Silver-loaded paint can be purchased from paint or chemical supply houses, and comes in quite a few formulations. Some can be painted on and air dried. Some can be fired, or even be put on with a silk screen process.

For ceramics, the fired-on variety is useful. After firing, the surface can either be electrolytic plated, provided there is a contact for the current, or electroless plated with nickel, copper, gold or even tin.

An idea for making decorative jewelery is to dry leaves from trees or use clean seashells, paint with silver-loaded paint, air dry, and bright copper electroplate to 0.001in.

Also molded epoxy figures or busts can be made in a similar way and this is an inexpensive means of making ornaments.

CHAPTER 7

Electroless Electroplating

The reason for the title of this chapter is that the deposition of metals can be carried out without using an electrical current. No electrical equipment is involved in the actual electro-deposition. It is sometimes referred to as chemical plating, because the chemicals in the formulation effect the metallic deposition. The main constituents of the solutions are an aqueous solution of the chemical containing the metal to be deposited, and an aqueous solution containing a chemical reducing agent. These can be mixed together to form the plating solution, but before any deposition takes place, a catalyst must be present. In this process the component is the catalyst. No anodes are used. It is merely a tank containing the electroless solution, with or without agitation, set to the correct temperature, with the component immersed in the solution.

The most popular and useful electroless solution is for depositing nickel. Usually this is deposited as an alloy of nickel and phosphorus, approximately 12% phosphorus.

ELECTROLESS NICKEL
Make up:-
Nickel Chloride Crystals. 5 oz. per gallon
30grms per liter.
Sodium Hypophosphite 1.5 – 1.6oz per gallon 10grms per liter.
Sodium Acetate Crystals 8oz. per gallon 50 grms per liter.

The sodium hypophosphite is the reducing agent in the solution. The best method for using this solution is to make up a solution of the nickel chloride and keep in one glass bottle, and make up a solution of the sodium hypophosphite and sodium hydroxyacetate together, and keep in another glass bottle. These can be stock solutions. When required, enough volume is mixed to plate the component. The ideal container to plate with this solution is a heat resistant glass beaker, heated by a Bunsen burner, gas ring or electric hot plate.

The operating temperature for this solution is 88°C–94°C (185°F – 200°F).

Some agitation is required, but an occasional shake of the wired component will release the bubbles of hydrogen that collect on the component as a product of the reduction process.

The deposition rate for the solution at 88°C (185°F) is:-
0.0006in. (15 microns) in sixty minutes.

However, to maintain this rate of deposition, small additions of the stock solutions of nickel chloride and the sodium hypophosphite with sodium hydroxyacetate have to be added at intervals, to keep the balance of the solution while plating. p.H should be maintained between 4.0 and 6.0 by additions of aqueous sodium hydroxide.

Because there is no electrolyte action, i.e. anode with outside electrical supply, there is no problem with 'throw', masking or high and low current density areas on the components. The deposit of nickel alloy is of even thickness all over the component. This process is useful for nickel alloy plating down narrow holes, orifices or tubes.

Ferrous metals can be plated in this solution after the pre-clean listed in chapter four. Copper and its alloys can also be plated, but, being non-catalytic, need to be touched with iron or aluminum wire to start the plating operation. Aluminum can be plated direct with this solution, no zincate dip being needed.

The metallurgical properties of the plating are interesting. As plated it has a hardness of 500 V.P.N. By heat treating up to 400°C the hardness can be increased to 900 V.P.N. The appearance of the deposit can vary from dull to semi-bright metal.

An example of mild steel components, electroless nickel plated. This deposit will give an even thickness over all the surface of the component, however complicated the shape.

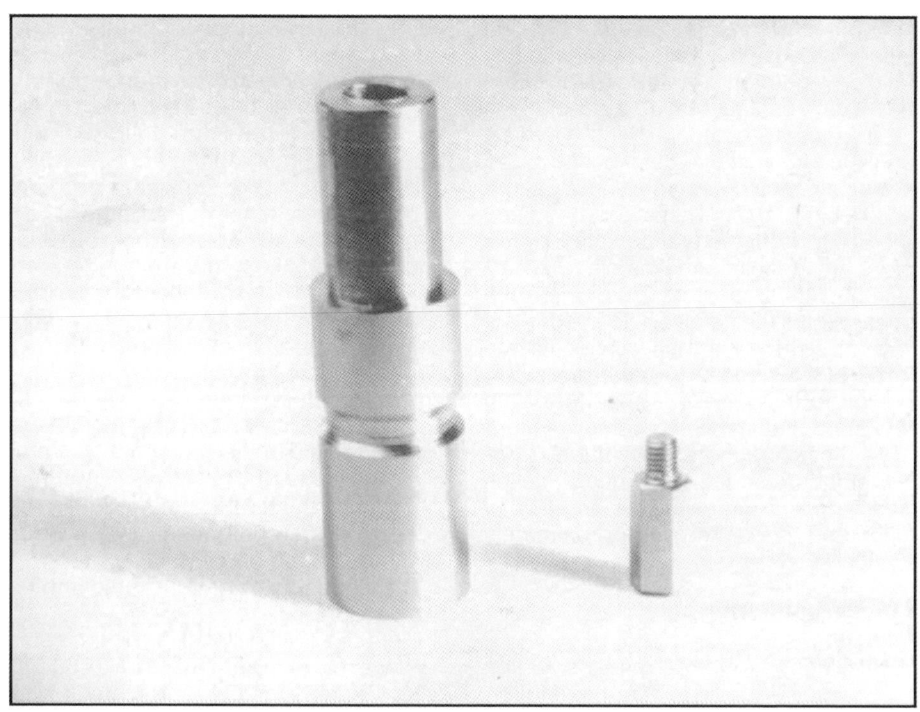

The electroplating supply houses offer an excellent range of electroless nickels.

TIN ELECTROLESS PLATING

The following two formula are useful for putting a thin layer of tin over ferrous metal and copper and its alloys.

Depositing tin or copper by this method makes it useful for soldering, especially on printed circuits, where there is no common connection for electrolytic tin electroplating.

For depositing on copper and its alloys
Electroless Tin

Make up:-
Stannous Chloride 1.6oz per gallon 10grms per liter.
Thiourea 13.6oz per gallon 85grms per liter.
Concentrated Hydrochloric Acid 2.3fl.oz per gallon 15mls per liter.
Water to make up the volume.

The solution is used at a temperature of 50°C (122°F).

The usual immersion time is five minutes. Note: this formula is patented.

For a quick pre-clean on printed circuits, give a light rubbing with a slurry of scouring powder, rinse, and immerse in the solution.

For immersion coating tin on steel, the following solution is used.

For depositing on iron and steel
Electroless tin for iron and steel

Make up:-
Stannous Sulphate 0.24oz per gallon 1.5grms per liter.
Concentrated Sulphuric Acid 0.14fl.oz per gallon 4mls per liter.
Water to make up the volume.

Use the pre-clean for steel as set out in chapter four, then immerse in the tin solution at a temperature of 82°C (180°F).

The time of immersion will vary with the surface condition of the material. This could be between one and ten minutes.

Containers for immersion tin plating are heat-resistant glass, or plastic containers, usually polythene or propylene. For these containers the aquaria heaters can be used.

The thickness of deposit from electroless tin solution is thin.

CARE with the acids used – gloves, goggles and overalls must be worn.

The electroless tin solution for copper being rather solid, when heating from cold some 'bumping' will take place, until the solid chemicals are dissolved.

CHAPTER 8

An Example and the Consideration of Electroplating

All the chapters dealt with so far have been set down in a kind of logical sequence of the processes of electroplating. So, assuming the average reader is starting on a small scale, and may be a model engineer, he or she wants to electroplate a batch of small components.

Taking the hub caps of a small traction engine as an example:- The components have the shape and dimensions set out in Fig.15.

The components have been machined to a good surface finish, and are ready for electroplating, after being stored in a dry area. In other words, there is no heavy surface rust.

The next part of the operation is to calculate the surface area of the components to be electroplated. This can be done from the diagrams, or the parts measured by hand. Either way it is the same.

To calculate the area of the hub caps:-

1) For $1\frac{1}{8}$ in. dia the area is $\pi r^2 = \dfrac{22}{7} \times \dfrac{9^2}{16}$

$\qquad\qquad\qquad\qquad\qquad = \underline{0.99\ sq\ in}$

2) Taking circumference x height of $\frac{5}{16}$ in.

$2\,\pi \times r \times$ height

$2\,\pi \times \frac{3}{8} \times \frac{5}{16}$ $\qquad\qquad\qquad = \underline{0.72\ sq\ in}$

Total area is $\qquad\qquad = (0.99 + 0.72) \times 2$

Both inside and outside $\qquad = \underline{3.42\ sq\ in.}$

For four hub caps $\qquad\qquad = 13.68\ sq\ in.$

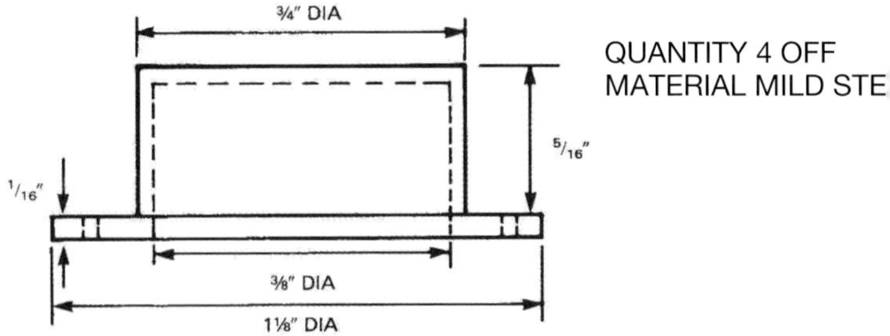

Fig. 15 *Hub Caps*

The hub caps being made of mild steel, the electroplating will be 0.0001in. of Semi Bright Nickel and 0.0005in. of Semi Bright Copper.

The current density in both electrolytes will be 10a.s.f., (see chapter five). Therefore, for nickel and copper the electroplating current will be:-

$$\frac{10 \times 13.68}{144} = 0.95 \text{ amperes.}$$

Remembering one square foot = 144 square inches.

The time for the electroplating for nickel = 18 minutes.

The time for the electroplating for copper = 60 minutes. (See chapter five).

That completes the theoretical part of the electroplating process, and now comes the practical part.

The hub caps are buffed to the desired luster with a mop or other polishing tool. After polishing, the components are immersed in an organic solvent, e.g. white spirit, for five minutes, then wiped with a rag soaked in the spirit. The components are then air dried. They are then hung via one of the small screw holes on copper wire at the required length to fit the depth of the cleaners and the electrolyte.

After wiring, immerse back in the white spirit for thirty seconds, and allow to air dry. This is to make sure there are no finger marks.

Being made of mild steel, the cleaning cycle is as follows, from the tables in chapter four.

(1)   Immerse in alkaline soak cleaner at 80°C (176°F) for three minutes.

(2)   Immerse in alkaline electrolytic cleaner at 80°C (176°F), for one minute, with the components connected to the cathode ⊖, negative part of the rectifier. This gives a cathodic clean. (Watch for gassing on the components, this gives an indication of the cleaning effect). After one minute, reverse the polarity, i.e. connect the ⊕, positive rectifier clip to the components and anodic clean for about ten seconds.

After the alkali cleaning, rinse in cold water for thirty seconds and transfer to No. 1 or No. 2 pickle, immerse for one minute at room temperature.

Rinse in cold water for thirty seconds.

Connect the rectifier to the nickel tank, positive ⊕, to the nickel anodes, cathode ⊖, negative, to the cathode bar. Check the circuit by switching on the rectifier on small adjustment of current, and dip the clip from the cathode bar into the electrolyte. If current reading is obtained, the circuit is the correct polarity.

With the electrolyte at the required temperature of 50°C (130°F) and the rectifier set to approximately one ampere, connect the wired components to the cathode bar, immersing in the electrolyte at the same time. After the components have been wired onto the cathode bar the current is adjusted to 0.95 or 1.0 amperes.

The compressed air agitation is adjusted for a steady bubbling action, then left for eighteen minutes to electroplate.

When this time has elapsed, the wired components are removed and rinsed in cold water.

While they are rinsing connect the rectifier to the copper electrolyte (the same procedure as for the nickel electrolyte). The circuit polarity is checked (see nickel electrolyte). With the electrolyte at the required temperature of 23°C (74°F), and the rectifier set at one amp, the wired components are removed from the rinse tank and connected to the cathode bar, immersing in the electrolyte at the same time.

When all the components have been wired onto the cathode bar, adjust the current to 0.95 or 1.0 amp. The compressed air is adjusted for a steady bubbling action and left for sixty minutes.

After sixty minutes the components are removed and rinsed in cold water, then in hot water. They are then dried off in hot air.

After drying, the components are unwired, and lightly rubbed with mutton cloth or similar soft rag. If needed, a light buff enhances the appearance.

CONSIDERATIONS IN ELECTROPLATING

This is by way of a final word on some of the aspects mentioned, but not elaborated on in various chapters of the book.

Mention is made of stripping the electroplated deposits for faulty appearance or bad adhesion. There follow some metal stripper formulations:

1). **For stripping nickel from ferrous materials, and copper and its alloys:- Nickel stripper.**
Make up:
Three parts *concentrated sulphuric acid.* Two parts *water.* by volume.

The stripper is used at room temperature, with the components anodic, positive $\oplus$, and using lead strips as cathodes, $\ominus$, negative. Voltage 4 – 6 volts. If required, 3oz. per gallon of glycerin may be added. This prevents etching, especially on steel. The nickel-plated components are left in the stripper until the nickel is completely dissolved off, showing the substrate. If left too long

in the stripper the substrate will start to dissolve away.

2). **For stripping copper deposits from steel and copper and its alloys.**
Use the recommended nickel stripper, (No 1).

3). **For stripping of copper with nickel undercoat.**
Use the recommended nickel stripper, (No 1).

4). **For stripping nickel from aluminum.**
Immerse in 50% aqueous nitric acid, or concentrated nitric acid at room temperature, until the nickel deposit is dissolved off the aluminum.

5). **For stripping anodize.**
The following solution can be used:-
Make up:-
Concentrated sulphuric acid. 16fl.oz. 100mls per liter.
Potassium Fluoride. 6 oz. 40grms per liter.
Water. to make up 1 gallon (to make 1 liter.)
The solution is made up by adding water to a heat resistant glass or plastic tank (polythene type). The sulphuric acid is added slowly, and stirred continuously. After the addition of the acid, add the potassium fluoride, stirring continuously until dissolved. Allow to cool to room temperature. Adjust to final volume with water. Use the stripper at room temperature. Allow the components to stand in the stripper until the anodize is dissolved off. The information on stripping defective anodize in chapter nine is still relevant, but a separate stripper may be needed, especially with dyed parts, which would color the anodize pre-treatment. This gives a choice.

6). **For stripping tin from steel and**

copper and its alloys.
The following solution can be used:-
Make up:-
Copper sulphate crystals. 8 oz. per gallon 50grms per liter.
Concentrated sulphuric acid. 16fl.oz.per gallon 100mls per liter.
Water. to make up 1 gallon (to make 1 liter).
The solution is used at room temperature.

All the listed strippers can be used in plastic containers, like polythene. *CARE* with sulphuric and nitric acids, both are corrosive. Use goggles, gloves and overalls. Add acid to water, not vice versa. Watch for heat generated by the reaction. After each part of the processing, i.e. between the cleaners, pickle, bright dips and the electrolytes, a cold water rinse must be carried out.

This cleans the components, and stops the reaction, and prevents carry-over of the solutions, thus preventing contamination.

All of the formulations have excluded all the cyanides and chromic acid. They have been mentioned in the text, but the reason for excluding these electrolytes, such as silver and gold, is because they contain cyanide. You might be able to purchase them from the various supply houses, but scheduled poisons are only made available to genuine industrial electroplaters.

The exclusion of chromium electroplating is because chromic acid is used in the process. In industrial electroplating mandatory regulations apply, and the hazards with chromium electroplating, when carried out in an unregulated area, can be very considerable.

USEFUL INFORMATION

To convert °F to °C	(Temp °F − 32) x $5/9$
To convert °C to °F	(Temp °C x $9/5$) + 32
To convert ounces per gallon to grams per liter	multiply by 6.25
To convert grams per liter to ounces per gallon	multiply by 0.16
To convert fluid ounces to milliliters or cc	multiply by 28.35
Nickel Sulphate Crystals	Formula $NiSO_4\ 6H_2O$
Nickel Chloride	$NiCl_2 6H_2O$
Sodium Hydroxide (Caustic Soda)	$NaOH$
Potassium Hydroxide (Caustic Potash)	KOH
Sodium Carbonate (Soda Ash)	Na_2CO_3
Concentrated Sulphuric Acid (SG 1.84)	H_2SO_4
Concentrated Nitric Acid S.G (1.38)	HNO_3
Hydrochloric Acid	HCl
Copper Sulphate Crystals	$CuSO_4 5H_2O$
Hydrogen Peroxide	H_2O_2

CHAPTER 9

The Finishing of Aluminum and its Alloys

The finishing of aluminum can be accomplished by either an anodizing process or a conversion coating.

The anodizing process. This can be divided into sub-processes:-

SULPHURIC ANODIZE AND CHROMIC ACID ANODIZE

Sulphuric acid anodize is the more practical process for the amateur and model engineer. Most grades of aluminum and aluminum alloys can be anodized by this process; the purer the aluminum the better the anodized film. The various aluminum alloy constituents, i.e. silicon and manganese, tend to retard the process of the anodized film, either in the pre-treatment or the actual anodizing.

Most fabrications from sheet or extrusions and some castings can be successfully anodized. However, some alloys and some castings are not conducive to the anodizing process. Castings are usually anodized by the chromic acid process.

The process of anodizing is the laying down or the growing of a film of aluminum oxide over the surface of the aluminum. Aluminum forms a layer of the oxide very quickly, even after a chemical clean, but the thickness of the layer is dependent on time. The anodizing process accelerates the time, and also gives a denser unbroken layer of oxide, which enhances the properties of aluminum against corrosion. In the sulphuric acid process it adds a mordant layer which can be used for the dyeing or coloring of aluminum. However, for coloring the main essential is to have a good layer of aluminum oxide. Magnified 1000 x the layer would look like Fig. 16.

The important characteristics are the barrier layer and the pore size. The thickness of the barrier layer is proportional to the voltage. The pore size is dependent on the concentration of the electrolyte (sulphuric acid), the electrolyte temperature and the applied current.

Two factors emerge which make for a particular anodizing film.

Lower concentration, i.e. 10% sulphuric acid, and low temperature give a small pore size but produce a hard anodized film.

Higher concentration, i.e. 15% sulphuric acid, and higher temperature give a larger pore size which produces a film suitable for dyeing.

A good compromise is 12% v/v sulphuric acid which makes a good general anodizing bath.

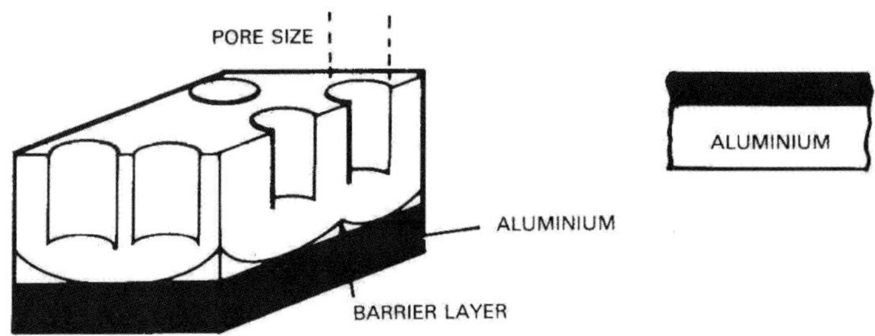

Fig. 16

Solution Preparation

To make up an anodizing bath a glass or a polythene type plastic tank is recommended. A good tank for doing small components is a large ice cream container. These have a capacity of about 1 gallon (4.5 liters). Half fill the tank with distilled or deionized water, and then slowly add the necessary volume of sulphuric acid, constantly stirring. *CARE!* The acid must be added to the water and not vice versa. It is advisable to wear protective gloves and safety glasses. After allowing the solution to cool the volume is corrected by the addition of more distilled water. An example of the dilution is as follows:-

10% volume solution for hard anodizing.
450 mls per 4.5 liters.
16 fluid oz. per gallon.
OR
12% volume solution for general anodizing.
540 mls per 4.5 liters.
19.2 fluid oz. per gallon.

For ease and convenience battery acid sold at garages can be used. This is of a dilution of 33% of sulphuric acid in distilled water, approximate specific gravity of 1.275/1.280.

If the battery acid is diluted 1 volume to 2.0 volumes of distilled water, this would give a 10% sulphuric acid concentration.

If the battery acid is diluted 1 volume to 1.5 volumes of distilled water, this would give a 12% sulphuric acid concentration.

For a gallon solution the dilution is 3 pints of battery acid to 4^{1}/2 pints of distilled water, giving a total of 7½ pints total volume. Just under the gallon. If measuring in liters or mls. the dilution is 2 liters or 2000 mls. to 3 liters or 3000 mls., giving a total volume of 5 liters or 5000 mls., which is just over the 4.5 liters; the equivalent to a gallon.

Setting up of the anodizing tank

After the required volume of solution of sulphuric acid is made up (either 10% or 12%) two pieces of clean sheet lead are placed down two opposite sides of the tank. These are bent over the top of the tank to position them and to make them convenient to connect to the electrical supply. These are the cathodes. The anode bar is placed down the middle, and the components to be anodized are connected to this. The usual way to connect the aluminum components for anodizing is to jig them in a titanium made sprung jig. These come in different shapes for various components, but are expensive to buy. Titanium rod and

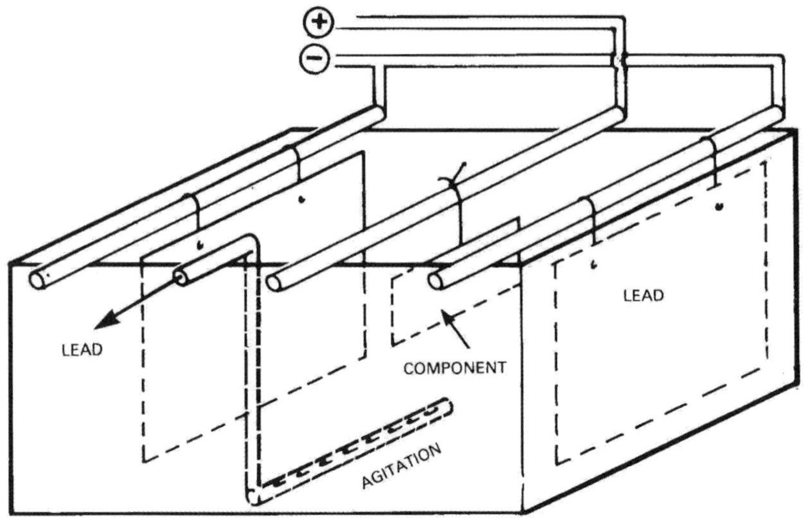

Fig. 17

small sized bar can be purchased for making these jigs. These can be used for anodizing a large quantity of similar components, which would prove the most economical way of processing them.

Generally the use of aluminum wire is the most economical for small quantities, and it can be bought in various diameter sized reels of varying stiffness. Lengths are cut slightly longer than required and coiled over a bar of metal or wood so that some tension is obtained. The component is wired up using a convenient hole, or a part of the surface that will secure a good electrical contact. The wire is then twisted to ensure a rigid contact. The rest of the components are wired the same, if you are anodizing a batch or a set of components. The wired components should hang approximately midway in the electrolyte, with an allowance on the wire for attaching it on the anode bar. The wires are then snipped off to that size. This will reduce the surplus wire and avoid the possibility of shorting on the cathode bar

as well as making it easier and neater to wire onto the anode bar.

With the components wired up, the power is switched on and adjusted to the required current density for anodizing. It is then left for the required time to obtain the required thickness of anodize.

To enhance and improve the quality of anodized finish, agitation is recommended, but good quality anodize can be obtained in still baths. To agitate the electrolyte a plastic pipe with holes drilled in is connected to a flexible pipe (Fig.17), and in turn connected to a regulated compressor. This will blow air into the electrolyte causing movement by bubbling. Care must be exercised in regulating the air flow, or the electrolyte will bubble over, and could prove dangerous and cause accidents.

Operating Conditions

The plating current for anodizing should be between 10 – 15 amps per square foot, or 0.069amps per square inch to 0.104amps per square inch of the surface area to be anodized. The voltage is to be between 12 and 20 volts. At 15 amps per square foot the deposit of anodize will be 0.0001 in. in 6.5 minutes. However, this will take longer depending on the amount of agitation. With no agitation it will take 10 minutes to deposit 0.0001in. of anodize. The temperature of the process is between 20 – 25 degrees C, 70 – 75 degrees F. If the electrolyte exceeds the top limit allow to cool.

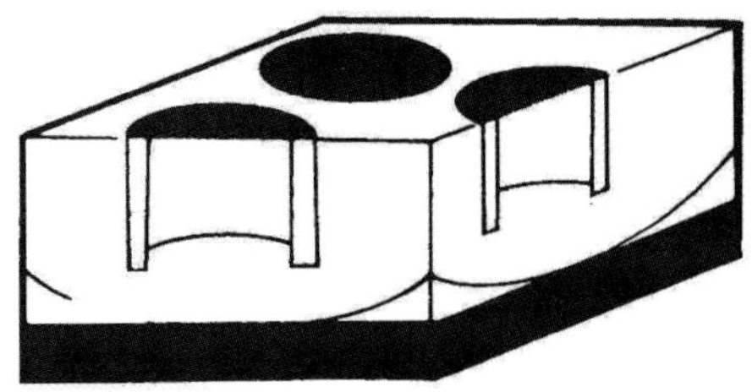

Fig. 18 *Sealed pore sites.*

Sealing of the Anodized Film

If desired the anodized film can be sealed. This process in effect seals over the pore sites of the grown aluminum oxide.

This is brought about by an increase in temperature in a sealant, which increases the volume and swells the cell walls and closes the pores.

The sealant is usually boiling deionized water a temperature of 100 degrees C for a period of 20 minutes. It is best done in a stainless steel container. An alternative is to boil pure deionized water in a stainless steel container with a lid, and suspend the anodized article in the steam given off for 30 minutes. The sealed articles are taken out and allowed to dry in the air, or in a convenient oven.

THE DYEING OF ANODIZED ALUMINUM

The majority of aluminum alloys when anodized have an appearance of natural aluminum or 'silver' color. Some alloys contain alloy constituents such as manganese, silicon, magnesium and copper. If these materials are present between 5%-8%, they produce a coloration of the anodized film. The coloration varies from a brown for manganese, for silicon and magnesium a bluish gray, and for copper an orange yellow.

To obtain a consistent reproducible uniform color, the easiest and most economical way is to use an organic dye that is soluble in water. To obtain the required shade of color various concentrations of the dyes have to be used. A guide is between 1 grm.per liter to 10grms per liter, O.16ozs.per gallon to 1.6ozs.per gallon. This concentration will also vary with color.

Operating Conditions for Dye Tanks

The ideal tank for dyeing should be made of good quality stainless steel, Austenitic grade 320S17. However, most plastic or glass tanks can also be used, but they will all require heating to approximately 40 – 50 degrees C, 104 – 122 degrees F.

The dye is weighed out according to the concentration, and added to half the volume of hot water contained in the dye tank. The solution is agitated by stirring or by air agitation from a regulated compressor. This is continued until all the dye has dissolved and no particles remain in suspension. Suspended particles of dye are the main cause of streaks on the dyed surfaces of the components. After dissolution of the dye, the solution is made up to working volume, and the desired temperature maintained by the heater.

After thoroughly rinsing the components in cold water they are hung across a bar of metal or plastic and totally immersed in the dye, preferably with agitation for the required time. The time, being dependent on the intensity of the color, is usually between 5 and 15 minutes. To maintain the consistency of color for subsequent batches of components, the conditions of temperature and time must be closely adhered to.

A flow process chart for the various processes in anodizing is shown in Fig. 19.

Points to remember for good practice in the anodizing process are:-

(a)   Thorough cleaning in the pretreatment cycle, and the complete removal of grease and smut from the components.

(b)   Obtain, and maintain, a good electrical contact between the wire on the component and the anode bar.

A point to remember is that the anodized film acts as an electrically insulating surface, so if the contact moves, the conducting surface under the wire in the electrolyte is lost. The consequence of this is that the growth of anodized film stops, but the worst part of this action is that the resistance at that point increases the heating effect, causing erosion on the surface.

(c)   After the anodize part of the process, the components are removed after switching off at the rectifier, and quickly rinsed in cold water, then quickly immersed either in the dye bath or the sealing tank (for natural anodize). At no time should the anodized film be allowed to dry. To do so would give a partial closure of the pore area, and hence restrict the ingress of the dye.

(d)   A good method of checking anodize is to use a test meter (Avo type) set on D.C. low voltage, and

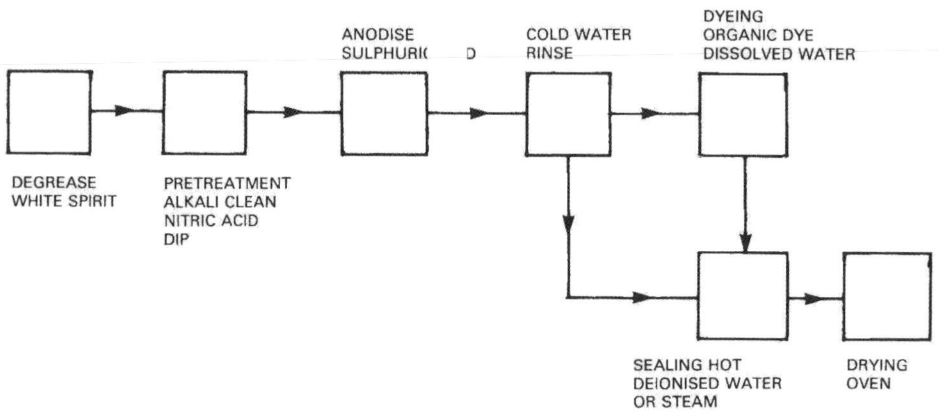

Fig. 19

An aluminum component anodized and dyed black following the procedures described.

set to resistance. Lightly pass the meter probes across the surface. Any areas with no anodize will conduct, which will cause a deflection on the meter. A simple check for a sealed surface on natural anodize is to moisten a tiny area of the surface, and touch the surface with an indelible pencil. If the surface is then wiped clean, leaving no trace of the color of the pencil, the surface has been satisfactorily sealed.

(e) To strip defective anodize from components, the wired components are immersed in the alkali cleaner for a few minutes, rinsed in cold water, immersed in the nitric acid solution for a few minutes, rinsed and re-anodized. In other words, back through the pretreatments.

Index